PHYSICS, CONSCIOUSNESS
AND
THE NATURE OF EXISTENCE

by

Joseph Norwood

ISBN: 1-4033-3426-9

This book is printed on acid free paper.

1stBooks - rev. 04/05/02

DEDICATION

To Sir Arthur Stanley Eddington

He intuited the answer before anyone else was even aware that there was a question.

PREFACE

In which we speak of some matters not
usually discussed and advance the notion
that wisdom has practical value.

Man's intelligence, together with the manual dexterity afforded by his body
and the stimulation of the challenges offered by his environment have allowed
him to evolve to a dominant position in his own little corner of the Universe.
Almost every aspect of Man's assent has been examined, analyzed, and dissected
by experts working in various fields and disciplines. All the facts are there in this
vast heap of disparate information, but no one in the physical or social sciences
really has a broad enough overview to understand the overall nature of existence
and the meaning of life - who and what we are and where we are going. The
question bridges too many disciplines.

As our knowledge about ourselves and our world has increased, the task of
the person who seeks to have such a panoramic overview has grown more
difficult, largely owing to the seemingly endless tendency toward specialization.
The reason for this tendency seems obvious. The sheer mass of Man's knowledge
is overwhelming and is much more than any one person could reasonably aspire
to absorb. This does not entirely explain the nervous smirks that greet the Big
Questions or the wall of suspicion and indifference that has grown up between
the various branches of human knowledge. Anyone who has ever served time in
the academic world will know what I mean. Physics and psychology have
nothing to say to each other. In today's universities where one is usually trained
rather than educated, everyone who is not a philosopher tends to regard
philosophy as a useless subject in the modern world. Intellectuals either disregard
religion altogether or, if they are religious, feel as though they must wear one hat
and brain set on Sunday and an entirely different one for the weekdays.

There is no shortage of experts on the various trees and shrubs that make up
the forest, but there is almost no one who knows what the forest as a whole looks
like. Even in philosophy, which might be expected to be concerned with such
questions, metaphysics is not very fashionable. Does this really matter? Yes, it
does.

If you have a problem, you will seek out the advice of a recommended
specialist in the area of your problem, not some omniscient sage, even assuming
you could find one. But does success in life really consist of solving problems?
Wouldn't it be much better not to have problems in the first place? Solving
problems takes knowledge. Avoiding problems takes wisdom, which is an
altogether rarer commodity than knowledge. Surmounting problems, one after

another, is wearing. You conquer one dragon, briefly savor the victory, and then another one is on you. The acquision of wisdom is the only way off of that reactive treadmill, and for this reason alone, wisdom has more practical value than you could possibly imagine.

Human beings are very neurotic, and almost all of us settle for this painful condition as normal and inescapable. Most people concentrate on entertaining themselves and run from knowledge (much less wisdom), knee-jerking their way through life, complaining about the unfairness of it all and refusing to take responsibility for themselves. The bumper sticker, "Life's a Bitch and then you Die", pretty well says it all. There are a few people who are born curious and succeed in gaining some knowledge, even deep knowledge. However even this elite behave in a manner that is totally at odds with that which they know. As this is being written, our paradigm, the basic instinctual assumptions about the nature of existence, the common sense that shapes our lives, is almost 100 years behind our knowledge. We still behave as though we lived in the deterministic, objective, mechanistic world described by classical physics. To be sure, appearances are what we deal with day in and day out, but if we look at the inescapable implications of twentieth-century physics, we see clearly that those appearances and the paradigm based on them are an illusion, and an unhealthy illusion at that.

The new physics, quantum mechanics, is an outstandingly successful theory, but it lacks the underlying principles and conceptual foundations that we were used to in classical physics and which gave us a comfortable feeling for how 'the little men were running around down there'. The impressions of the underlying reality behind the new physics appear chaotic to those few who have bothered to look. Nothing is more unnerving than chaos. The fear of chaos is the most visceral fear there is and it is the combating of this fear that was the main original function of religion, and one with which religion is still concerned today. Consequently, the rule in physics today is, 'calculate and measure, but don't think', which is, as we shall see, not very good advice.

Science, which was given birth by the Church, came to compete with it for the soul of Man. In the West, this contest was particularly bitter. As the age of whatever miracles there had been receded further into the past, the Church came to insist upon blind faith in its dogmas. To science, blind faith was a coin of no worth whatsoever. Looked upon as a contest of who could bring home the bacon, science won. Consequently, most of us live desperate and spiritually bereft lives in an age of dazzling technological achievements.

Philosophy always had one foot in the Church and was, to some extent, also a partner of science up to fairly recent times. In the nineteenth century, all well-educated men were expected to be conversant with philosophy and scientists were usually referred to as natural philosophers. The terminal degree in physics and other academic disciplines, the Ph.D., the Doctor of Philosophy, dates from those times. As the twentieth century dawned, physics began to delve into areas further removed from common experience and to come up with answers that were not at all comfortable to contemplate from a philosophical point of view. So physics continued to award doctorates of philosophy, but philosophy and philosophers came to play a role in physics' scheme that has diminished to zero.

In the West, psychology got a later start than it did in the East. It attempted to adopt for its use the scientific method that had been so outstandingly successful for physics, but the human mind is subtle and complex and the connection there between theory and experiment is not at all straightforward. In the East, practical psychological methods became the core of the esoteric approach to religion, and it was this psychological as opposed to philosophical approach that distinguishes Eastern esoteric religion from the Western exoteric religious tradition. It seems that there may now be the basis for a partnership between physics and psychology, as we shall see. Recent developments also point to a need for physics to reopen the door and invite philosophy back in. A partnership with religion looks as though it might also be quite fruitful and beneficial to the aims of both parties. The opportunities are there and the rewards of such a synthesis would, I am sure, exceed our most optimistic expectations.

In writing this book, I have drawn on a vast amount of work by people who never spoke to each other. In this book, they will enter a dialogue, like it or not. It will be my task to draw meaning from this dialogue and to paint the forest for you in all its glory.

All the facts that you will need to draw some very useful conclusions are here. I must be up front with you, though. You are going to have to exercise your mind and be prepared to question your assumptions. Chapters Two through Twelve deal with physics. That is where all of the clues are to be found and you will have to understand them rather thoroughly in order to find the argument convincing. It will get easier as you go along, and for science buffs, I shouldn't think that there will be any problem. Those of you who are more philosophically inclined might read Chapter One, and then Thirteen through Eighteen to cut straight to the conclusion. Hopefully, this track will give you the incentive to then go back and deal with chapters Two through Twelve.

If this was just another book on modern physics, I would not have written it. There are a number of good ones already in print. In the last quarter of the twentieth century, several books have appeared written by people who were struck by the similarities and parallels between modern physics and Eastern esoteric religion. None of these books managed to draw any convincing conclusions from their observations. Another group of books has emerged more recently, written by people who have renounced the conventional interpretation of quantum mechanics and have laid claim to a realist philosophical position. They are willing to go to great lengths to preserve the deterministic paradigm and the existence of a real material world out there. A number of guesses as to the basic nature of this reality have been put forward, some of them quite bizarre, but they all agree on a materialistic metaphysics. The psychological reason for this assumption is not hard to find. If the world is a mental creation as several eminent philosophers have suggested, then the physicists would be made to feel as though their work in dealing with an 'illusion' had been totally devalued. This is nothing more than a concept, but it is a generally held position none the less. The upshot is that almost no one, with some few eminent exceptions, seems to have understood the conceptual underpinnings of twentieth-century physics. It is my aim in this book to correct this omission in such a way as to make this revelation available to a wider readership. If I am correct, and I think that I am, the message there is such an uplifting one that the story must be told. I can only hope that I have done it in such a way that you, the reader, will be stimulated spiritually as well as intellectually.

Fernandina Beach, Florida   JN

CONTENTS

PREFACE - In which we speak of some matters not usually discussed and advance the notion that wisdom has practical value. v

1- THE EMERGING MIND OF MAN - in which we trace the development of Man's consciousness in its spiritual and intellectual aspects. 1

2- THE CLOCKWORK UNIVERSE - in which we examine the classical physics of 'stuff' and its implied view of the world. 7

3- THE GHOST IN THE MACHINE - in which we look at the classical physics of 'not stuff' 17

4- CRACKS IN THE FOUNDATION - in which we pay the price for consistency between our theories of 'stuff' and 'not stuff' 22

5- THE FUN-HOUSE MIRROR - in which we examine Einstein's General Theory of Relativity where space-time is curved and twisted by its contents 34

6- THE MACHINE BEGINS TO CRUMBLE - in which the grim reaper knocks on the door of classical physics. 51

7- THE MADNESS GETS WORSE - in which several young men and a couple of older ones conspire to do away with determinism 68

8- AND YOU THOUGHT YOU HAD HEARD THE WORST – in which an analysis of measurement consigns objectivity and locality to the boneyard along with determinism 77

9- THE GOSPEL ACCORDING TO ST. NIELS - In which we begin to suspect that there is something rotten in the state of Denmark.............. 87

10- BOOLE GOES DOWN FOR THE COUNT - in which we must make the choice between physical weirdness and quantum weirdness................ 94

11- **SURELY ONE AND ONE STILL MAKE TWO -** in which the phrase "staying in touch" takes on a whole new meaning 102

12- **THINGS THAT GO BUMP IN THE NIGHT -** in which we cautiously dip a toe into the murky waters of parapsychology in search of the quantum connection 110

13- **REALLY, JUST FOR ME? -** in which we appreciate just how well suited for our comfort this Universe is 114

14- **METAPHYSICS IS NOT A FOUR-LETTER WORD -** in which we come to some conclusions concerning the nature of existence 120

15- **IT'S ALL IN YOUR MIND -** in which psychology is invited to join in the study of consciousness and its powers to manipulate wave functions .. 124

16- **A RAY OF SANITY -** in which the scientific and spiritual disciplines enter into a mutually beneficial relationship 129

17- **WORDS CORRUPT -** in which we delineate the virtues and limitations of language, both verbal and mathmatical 134

18- **YOU CAN LEAD A HORSE TO WATER, BUT -** In which we contemplate some radical brain surgery 137

INDEX 141

CHAPTER ONE

THE EMERGING MIND OF MAN

in which we trace the development of
Man's consciousness in its spiritual and
intellectual aspects.

It is my optimistic contention that the view that will be revealed in this book
will become generally obvious, hopefully sooner rather than later, and that this
radical change in the way we perceive our world and each other will amount to a
cultural paradigm shift. On the basis of this admittedly ambitious assumption, it
is useful to go back to the dawn of Man's emergence and look at earlier paradigm
shifts as the story of mankind has unfolded.

Man developed a spiritual life long before he developed an intellectual life,
and so we will begin our journey there. The religion of primitive man was not
just one facet of his life; it was essentially the whole of it. Primitive religion bore
no resemblance to religion as we might think of it today. There was no morality
and certainly no idea of God or gods. Early man made no conceptual distinction
between the inner and outer worlds. The subject-object distinction that seems so
obvious to us was entirely unknown to our remote ancestors. Primitive man's
instinctual concern was with resisting the seemingly arbitrary forces of nature
and the uncontrolled psychic forces within himself. He yearned for security, just
as we do, but he dealt with a different set of problems than those with which we
are normally concerned. We can understand the fears of being eaten by a cave
bear, but we have much less of a handle on the internal psychic forces that early
man sought to control. For us, these have been pushed into our unconscious mind
by the intellect, and they only rarely emerge in dreams or in psychotic episodes.
These psychic events constituted the most vivid features of reality for primitive
man and were projected into the world around him. He did not categorize or
conceptualize. Mental projections, feelings, and natural objects all existed on an
equal plane for him, and everything was animated and related in manifold ways.
The world of primitive man was highly interconnected, which is to say, very non-
linear.

In such a chaotic environment, survival depends on developing an ability to
limit and direct these multitudinous relationships so as not to be overwhelmed by
them. This need gave rise to language, whose initial and primary purpose was to
fulfill this protective function. If one can name an experience, then one gains a
degree of control over it. In order to see what I mean, imagine yourself to be

1

walking along a jungle path in the dark. You dimly perceive a coiled serpentine form in the path before you. Until you can identify it, that is to say, until you can name it, the power of the unknown menace must be overwhelming. As soon as you are able to label it as a vine, its power vanishes completely. In this sense, language and the word are very powerful magic. For early man, the label and the thing were one. If he could erase the word 'snake' and replace it with the word 'vine', then the hitherto real snake would vanish and be replaced by the harmless vine. From our point of view, we can appreciate the value of recognizing something we thought to be a snake to be a vine, but, of course, we see nothing magical in that. In bygone times, there was magic in it, and in primitive societies, this magic still persists.

Language is a two-edged sword. We have seen the power inherent in being able to label things. Language also drains power away because it tends to solidify reality. In an age of magic when the world was fluid and dream-like, man could tap into his psychic powers without the necessity to call upon any external aid. As language began to develop, primitive man came to rely on such stimuli as self-inflicted wounds and psychotropic plants to project himself into the mental states necessary to break through the walls raised by language and fluidize reality to the point where he could call forth his psychic energies. Trauma and magic mushrooms both work in much the same way. The insult to the body gives rise to a spurt of adrenaline, which is a chemical cousin of LSD and has similar effects, especially as regards altering one's time sense. But methods like these come at a price, and it is a rather high one. Psychotropic drugs, whereas they do open the doors, can also lead to disintegration of whatever personality there is. They were then and are now, quite dangerous.

The body and being of primitive man was thoroughly embedded in nature. The entire world, including himself was perceived to be in a continual flux of change. His own personal identity could in no way be disentangled from the environment. Gradually, as one of the earliest manifestations of the emerging subject-object duality inherent in language, man came to develop a tribal identity and began to see himself as an undifferentiated cog in the tribal machine. The distinction between my tribe and other tribes began to enter his awareness. Near the end of the Neolithic age, a further and more significant language-driven psychological revolution took place. Man's ego became gradually self-conscious and separated itself from the environment and from other members of the tribe. The subject-object distinction, inherent in language, became fully obvious as the internalized subject and externalized object, me in here, everything else out there. Adam had bitten the apple.

Thought now became increasingly autonomous and began to differentiate the mind from the rest of the human being and to strive for the cold detachment of abstraction. Until this revolution occurred, everyone was his own one-man religion and dealt directly with nature on his own. As the growing sophistication of language made magic a harder and riskier thing to do, specialists began to appear and each band or tribe came to depend upon a single shaman who assumed the role of dealing with the forces of nature for the benefit of the tribe and the profit of the shaman.

Another very important facet of the development of language is that as the forces of nature came to be named, they began to assume the characteristics of deities, and religion entered the age of theism. As man grew in self-awareness, he began to relegate his magic powers to the nature deities, originally conceptualized in man's own image. The power of the shaman was gradually transmuted from that of a magician in his own right into a priestly role as intercessory mediator with the gods, in whom the power now came to reside. In this way, the psychic forces discovered by man and aroused within himself came to be transferred into external agents of his own conception and hence beyond his control. Man must now beg his favors from the gods to whom he becomes subservient. Progress has its price.

Gradually, the pantheon became simplified as minor gods faded into the background and the chief god moved to the forefront as God with a capital G. This God, by whatever name he is known - Jehovah, Allah, or Brahman - is the ultimate abstraction about whom nothing whatsoever can be known. The priests have completed their transmutation from the techniques of ecstasy that they practiced as shamans and have become non-ecstatic officiants of dogma and ritual.

From a spiritual point of view, man at this stage has reached the limits of his outward expansion into objective space with total power transferal to Brahman, the Ultimate Object, and he begins to turn back toward himself. This movement from exoteric religion toward esoteric religion began more than 3000 years ago in the East. The earliest hymns of the Rgveda go back at least that far. In the West, the esoteric movement of the Gnostic Christians was stamped out by about the third century CE and was ruthlessly repressed thereafter, emerging only in a cryptic form in the writings of the alchemists. Sufism and Qabbalism are the esoteric movements in Islam and Judaism. They have been somewhat better tolerated than esoteric Christianity, but only just, threatening, as they do, clerical authority. Interest in esoteric religion seems to be beginning anew in the West with the arrival of Zen in the 1950s and Tibetan Vajrayana Buddhism in the 1970s, however the foothold is tenuous.

The pressure on man to look within increases as he becomes more and more aware of his seemingly unrelievable neuroses and sufferings and his apparently permanent Soul, Ātman in the Indian context. Just as man had reached the limits of his world expansion with the notion of Brahman, the Ultimate exterior unknowable, so in search of the spiritual principle of unity within himself, man comes up against the limits of his conceptual faculties with Ātman, the Ultimate interior unknowable. Ātman and Brahman denote the infinity within and without, but only in terms of concept, not of experience, which has yet to tap its limits.

Exoteric religion has been driven by the philosophical principle of Eternalism. The alternative would seem to be Nihilism, and indeed the Western world seems to be made up of conventionally religious people, Eternalists, and atheists, Nihilists, and a sprinkling of agnostics, who simply haven't made up their minds. The reform that the Buddha introduced into the vedic tradition was to neither affirm nor deny Ātman or Brahman. Rather, he escaped altogether from the jungle of philosophical speculation surrounding the extremes of eternalism and nihilism by shifting the emphasis from philosophical speculation to practical psychology. He emphasized the psychological techniques and practices designed to uproot neuroses and suffering and allow the world to be experienced just as it is. These had been developed by the Vedic tradition but had been buried in a mass of philosophical speculation, which confused the issue for many people. The Buddha simply refused to deal with the philosophical issues, which probably explains why Buddhism died out in the land of its birth. Indians enjoy philosophy like Americans enjoy baseball.

This is a highly simplified version of man's spiritual history as I understand it. The tendency for the emergence of ego is initially a survival function that serves to fix the limits between self and not-self and protect man from a chaotic world. Having achieved this protective purpose, the ego tendency then becomes a negative trait because it promotes self-clinging and all the neuroses that arise from taking one's self to be singular, permanent, and independent. This is easily understood. Other, not-me, is not under my control and must be regarded as potentially hostile. In this way paranoia arises. When is it that we are most aware of our separateness from others? It is when the negative emotions of anger, pride, envy, fear, and greed arise. These fixations preclude the development of compassion and lead to isolation and mental stagnation. Thus, after self-affirmation has reached its zenith, further development is only possible through a reversal of direction and the establishment of harmony through negation, not of the ego, but of an erroneously conceived self. Please understand that this does not mean a return to the ignorance of Eden. Far from it. The Buddha, Lao-tse, and Christ were all aware of themselves as individuals, but they did not indulge in clinging to an erroneously conceived self seen as singular, permanent, and independent in a hostile world.

Now let's broaden our view to look at some of the mundane aspects of man's development. In man's earliest interaction with his physical environment he was a nomad, living in small mobile bands following the herds of game in their migrations and collecting fruits and berries as they became ripe. The science of astronomy had its beginnings in those early times. Only by carefully observing the heavens was man able to devise an increasingly accurate calendar and learn to navigate from one place to another. Our present highly concrete sense of space and time had its beginnings in that way. The tribe also depended on its shaman to find game through his clairvoyance, control the weather, cure diseases, and generally serve as the tribe's middleman with the spirits. The shaman was indispensable, but he was too spooky for social acceptance. I very much doubt that he was regarded as one of the boys. It is much more likely that he lived in isolation on the edge of the encampment.

Man's ability to give up the hunt and the round of nomadic wandering and settle in permanent towns and cities came about through the discovery of grains that could be cultivated, dried, and stored to provide imperishable sustenance through the winter.

As with other steps in man's development, there was a price to pay. Grubbing in the earth to raise crops may be much safer than hunting large game with primitive weapons, but it is certainly not as satisfying from a psychological point of view. One suspects that agriculture may have been invented by women, for whom it would have been an entirely positive development, supporting, as it does, the instinct for the security of a permanent home. Some people think that the story of Eve's temptation of Adam with the apple in the Garden of Eden is a metaphorical recording of this discovery of agriculture.

Whatever the psychological price, man now had the ability to build cities, accumulate surplus wealth, and develop a more sophisticated economic structure. The invention of writing, so necessary to passing knowledge accurately from one generation to the next, came about during this early stage of agriculture because of the need to keep warehousing records.

When people began to settle down in one place, the need for the shaman began to wane. With the food supply under control by mundane means, the age of magic, already weakened by the development of language, rapidly drew to a close and the function of the shaman was replaced by that of the priest. Thus the shift from nomad to agriculturist tended to coincide with the general shift of magic power from man to the gods.

Successful agriculture requires a very accurate calendar in most parts of the world. You have to know just when to plant and when to harvest. A week or two too early in planting and you might be hit with a freeze. In the worst case, the entire nation might starve or be driven back to the old ways, assuming they could remember them. The first astronomers were also priests, but the two offices gradually became separate. At the peak of the expansion of exoteric religion, the priests officiated over a structure of belief that depended critically on blind faith, the age of miracles having passed. On the other hand, the emerging astronomer-scientist found that his successes came about through careful observation and direct experience. This widening gap between orthodox exoteric religion and science led to an estrangement that we still see today, especially between fundamental Creationists and advocates of Darwinian evolution.

5

In the West, the Medieval period was a period of stagnation during which man was trapped between his worldly rulers and the Church, between whom all power was shared. This period was brought to a close in the mid fourteenth century by The Black Plagues that so drastically reduced the population of Europe and Asia. Western human consciousness was jolted into abandoning medieval superstition and embracing science as a means of controlling nature. The age of exploration and discovery and the flowering of the Renaissance was the positive result of this catastrophe.

The East, with its emphasis on esoteric religion, karma, and reincarnation was more passive in its response to the plague. Religion rather than science continued to dominate Eastern culture.

In the West, science has continued its expansion up to the present day. Within the last hundred years, our technology has enabled science to begin to make contact with the bedrock reality that underlies the world of appearances. In the next several chapters we will see how this came about.

So there we have a brief chronicle of the paradigm shifts that have marked the development of the human race. The beginnings of language, the realization of tribal identity, the emergence of ego, the agricultural revolution and the concurrent rise of priestly religion, and, most recently, the scientific revolution of the fifteenth century have all produced a rapid change in the way people see their world.

Let us now focus sharply on the development of Western science, specifically physics, and see if we can find the seed of the next great shift in our consciousness.

CHAPTER TWO

THE CLOCKWORK UNIVERSE

in which we examine the classical
physics of 'stuff' and its implied view of
the world.

The subject matter of the last chapter spanned a hundred thousand years or more. In this chapter and the next, we will largely be concerned with the period from the middle of the sixteenth century to the end of the nineteenth century. This is the period in which the scientific method was developed and in which most of classical physics was formulated. The rest of the book will focus exclusively on the twentieth century and on modern physics and its philosophical, psychological, and religious implications. Such is the measure of Man's exponential rate of acquisition of knowledge, doubling as it does every ten years or so. Would that wisdom could be acquired at such a pace.

We have divided pre-relativity classical physics into two parts: that concerned with the motion and interaction of matter (stuff), which is largely the concern of the science of mechanics; and the physics of not-stuff, which is the study of magnetism, electricity, and light.

Let us begin somewhat before the beginning, in Greece during its golden age. The Greeks considered scientific questions, but their methods were limited to speculation based largely upon aesthetic considerations. There was no attempt to verify these speculations by experiment. Even so, the aesthetic principle is not an entirely unreliable guide, then or now, and the early Greek philosophers and mathematicians managed to originate such ideas as matter conservation, inertia, the atomic structure of matter, and the finite speed of light.

Aristotle (384-322 BCE) began the trend away from a purely speculative approach to science. He held that the cosmos was *geocentric* and that the unmoving Earth was the point about which the Sun, Moon, planets, and fixed stars rotated in circular orbits. The idea of a fixed Earth at the center of the cosmos with Hell below and Heaven above suited the early Christian church, and they seized upon it and dogmatized the idea.

This sort of closed-mindedness reduces scientific inquiry to a form of primitive mysticism and was a negative development that Aristotle, a competent biologist, surely would not have approved. The dogmatic insistence of the church upon the geocentric model of the universe was to prove an insurmountable barrier to progress in this area for almost two thousand years.

By 150 CE, the Greek mathematician and astronomer Ptolemy in Alexandria was forced to complicate the geocentric model by assuming that the heavenly bodies moved on circles whose centers moved on circular orbits (epicycles) in order to preserve the basic features that the church insisted upon in the face of improved orbital observations.

Following the fall of Rome to the barbarians, Europe entered the well-named Dark Ages and the torch of scientific progress passed for a time to the Arabs. Alhazen (or, more accurately, Abu-'Ali Al-Hasan ibn Al-Haytham), born at Basra in the late tenth century, made notable discoveries in optics. He was the first to explain the apparent increase in the size of heavenly bodies near the horizon and correctly identified these effects with the refraction observed in a pool of water. His treatise on optics was translated into Latin in 1270, but did not circulate in Europe until its publication in 1572, by which time science was finally beginning to flower in Europe after a sleep of 1700 years.

A Polish astronomer, Nicholas Copernicus (1473-1543) became convinced that the orthodox geocentric view of the universe embodied in the complicated Ptolemaic model was wrong and that the Earth was a planet on the same footing as the others, all of them moving in orbits about the Sun. This very same theory had been advanced on aesthetic grounds by Aristarchus 2000 years earlier, but now its time had almost come. Copernicus found that the seasons and the observed retrograde motion of the planets could be easily explained on the basis of such a *heliocentric* scheme. Copernicus' theory was worked out in considerable detail and, owing to the invention of the telescope by Galileo soon thereafter, the Copernican theory was never successfully repressed. Copernicus, fearing persecution for heresy, withheld publication until he was on his deathbed and thus beyond the reach of the Inquisition.

Galileo Galilei (1564-1642) is generally known as the father of physics. He was born in the same year as William Shakespeare and died in the same year that Newton was born. His father was an accomplished musician and scholar and he urged Galileo to study for a career in medicine. At the age of 17, while studying medicine in Pisa, he discovered the principles of the pendulum by observing the oscillations of a large chandelier in the cathedral. By the age of 26, Galileo had drifted away from medicine and obtained an appointment as professor of mathematics at the University of Pisa. Galileo was very intelligent, but he was also arrogant. He began to dissect Aristotle's theories with a great lack of tact, illuminating errors as he found them.

Most of Galileo's colleagues refused to listen to his heresies against the church-sanctioned work of Aristotle, and a lifelong persecution of Galileo began. Galileo refused to be prudent as Copernicus had been. He was hounded out of Pisa in 1592 and moved to the University of Padua where he spent 18 years as professor of mathematics. He was fortunate to obtain such a prestigious post in a liberal environment, and things went well.

Padua was part of the Republic of Venice, which was dominated by mercantile interests. Professors were expected to produce useful and practical results. Galileo augmented his teaching salary by the sale of measuring instruments that he designed and built.

In 1608, during Galileo's tenure at Padua, a Dutch optician named Lippershey discovered, quite by accident, a way of arranging two spectacle lenses to make objects seem nearer but inverted. The news of this discovery reached Galileo in June of 1609. He worked out the physics involved in a burst of creativity and threw himself into the development of the telescope. By January of 1610, Galileo had built a 30-power refracting telescope with an erect image. This instrument had an optical arrangement rather like an opera glass. When Galileo trained his telescope on the heavens, a flood of wonders met his eye. The Milky Way was resolved into stars; the planets were seen as discs rather than points; the phases of Venus were revealed; mountains could be seen on the moon; and the moons of Jupiter were discovered.

Galileo was a skilled self-promoter. He wrote a little book called *The Starry Messenger* and dedicated the book to Cosimo di Medici, the Grand Duke of Tuscany. The telescope was put to practical use by installing it at the top of the Campanile, Venice's tallest tower, from which merchants could see incoming ships still several hours out, and set prices in advance. Galileo became rich and famous.

He then accepted the post of court philosopher and mathematician at the Medici court in Florence, where he was on public view and thus vulnerable to the enemies he had made at Pisa.

Galileo seemed to have taken leave of prudence in those years. In demonstrating his telescope before the Papal court in Rome, he claimed to have absolute proof that Copernicus' heliocentric theory was correct. The Pope was not amused and Galileo's growing number of enemies fanned the flames.

In 1615, he was summoned before the Pope and ordered in no uncertain terms to cease his blasphemies. Galileo had always been imprudent in sharing his theories and discoveries, but this meeting with the Pope apparently jolted him. He henceforth conducted his research in relative secrecy until 1623 when a friend and benefactor, Cardinal Barberini, became Pope Urban VIII. Galileo, who in addition to being arrogant was obviously more than a little naive when it came to politics, reasoned that this change at the Vatican would allow him to publish his theories openly.

Galileo began his great book of *Dialogues on the Ptolemaic and Copernican Systems*, which was published in 1632. Galileo did hedge his political bets to some extent. The book abided by the letter of the 1615 papal decree, but managed to make a devastating argument for the Copernican cosmology. In the book, three characters carry on a discussion: Saliviati, a Copernican; Simplicio, an Aristotelian; and Sagredo, an intelligent and impartial moderator. The dialogues cover four days; no conclusions are drawn, but the arguments speak decisively for themselves and the book made a solid case for the Copernican system.

Pope Urban's court was a hotbed of plotting and intrigue. The Pope became persuaded that Simplicio was a caricature of himself. Galileo was harassed for a time and was finally called before the Inquisition. He was, at this time, 67 years old and in failing health. He was shown the instruments of torture and condemned to close house arrest for the rest of his life.

Those last years of Galileo's life were, in many ways, his most productive. He was now out of the limelight and could concentrate his thoughts on the science of motion without ego-aggrandizing distractions.

Aristotle had regarded the speed with which things move as a contest between *propulsion* and *resistance*, a not unreasonable idea. Consideration of falling bodies, which were known to increase their speed, raised the obvious question of whether propulsion was increasing or resistance was decreasing. What role did weight play in this in light of the observation that light objects fall nearly as fast as heavy ones?

In 1636, Galileo published his *Dialogues on Two New Sciences* (cohesion and motion). With the dialogue taking place between Salviati (Galileo's mouthpiece), Simplicio (who argues most unconvincingly), and Sagredo (the intelligent pragmatist). With regard to falling bodies, Galileo concluded that all bodies of whatever weight will fall at the same rate in a medium devoid of resistance, that is, in a vacuum. He also stated that equal increments of speed are gained in equal intervals of time, which is to say that bodies fall in a vacuum with constant acceleration.

This is an extraordinary result in which Galileo has conceptualized a vacuum, which prevailing scientific thought deemed un-natural. He came to this conclusion by noting that two bodies that fall at vastly different rates in water will fall at more nearly the same rate in air. Thus he reasoned that in the hypothetical vacuum, the two rates would be equal. The man was a genius.

Galileo had been expressly forbidden to say that the Earth could move without our being aware of the motion, and so he let his observations on projectile motion make the argument for him. These observations were formulated as follows.

The Principle of Inertia states that an object moving at constant speed on a level surface will continue to move in the same direction unless it is disturbed.

The Principle of Superposition says that if an object is subjected to two separate influences, each known to produce a characteristic motion, then the object will respond to each influence without modifying its response to the other.

Consider the effect of these two principles on a person standing on the surface of a moving Earth. The first (inertia) allows you to share the motion of the Earth. Even though a point on the equator is moving at over a 1000 miles per hour (1600 kilometers per hour), if you jump up, you come straight down. The Earth does not race out from under you. The Principle of Superposition guarantees that your jump will respond to gravity just as though the Earth stood still.

A near contemporary of Galileo was the French philosopher René Descartes (1596-1650). Descartes did not want to give up the security of familiar Aristotelian ideas without having something solid and reassuring to put in their place. Thus he set himself the task of formulating a new general philosophy to harmonize the emerging science with theology. Descartes was a practical man, a civil engineer working for the Army. He set great store by logic and pure reasoning. His style is still very much in vogue in French academic circles. A basic principle is isolated and followed by impeccable deduction to conclusions of enormous scope and range. We see this Cartesian method at work in the stories of Sherlock Holmes, who was said to be French on his mother's side.

Descartes' mathematics was much better than Galileo's, but his physics was often flawed and confused. Much of the credit for Descartes' physics really belongs to his friend Christian Huygens (1629-1695), the son of a Dutch diplomat.

Descartes turned his attention onto the means by which one object transfers motion to another, and in so doing, he formulated the first physical *conservation* law. The conserved quantity is *momentum*, the product of mass and velocity. If you can identify some quantity such as momentum that does not change in the course of a complex process, such as a collision, then the solution of the problem becomes much easier.

Beside Galileo, a superstar in his own right, Descartes, and Huygens, there were a couple of other significant players who prepared the way for Newton. These were Tycho Brahe (1546-1601) and Johann Kepler (1571-1630). Their story is an excellent illustration of the necessary interaction between theory and experiment. Brahe, the experimentalist, was born into a noble family in Sweden and, working as a hobbiest, became the greatest astronomer of his time. In 1575, he was hired as court astrologer by King Frederick II of Denmark. Brahe was a huge, fierce man who had lost his nose in a duel and wore one fashioned of gold and silver in its place. His new employer gave Brahe the island of Hven, an observatory with state-of-the-art equipment, and a palace. The equipment consisted of long sighting tubes mounted on large circles graduated for precision measurement of astronomical position. Brahe worked there for 20 years, generating a marvelous catalogue of the thousand brightest stars and the most accurate log of continuous planetary positions ever taken. This work was done with great care and an amazing degree of accuracy was achieved in measuring the planetary orbits, all without a telescope. Galileo was not to build the first one for another 35 years.

Apparently Brahe's astrological predictions must have been satisfactory as well since history records a peaceful relationship between Brahe and his patron. In 1599, Frederick II died and his son, Christian II, who had other interests, turned off the money tap. Brahe left Denmark to establish a new observatory in Prague, entering the service of Rudolf II, the Holy Roman Emperor, who was himself an amateur astronomer. Brahe brought with him his Danish instruments and the precious tables of observations. He was sure that Aristotle had been right and that the Earth must stand still. He intended to prove it with his data. Brahe died in 1601 before the new work had progressed very far.

The mathematician Kepler was one of Brahe's assistants at Prague during the last months of Brahe's life. After Brahe's death, Kepler carried on with his calculations based on Brahe's planetary observations. In 1612, Kepler moved to the University of Linz and worked there until his death in 1630. Brahe had been a firm believer in the Aristotelian view of a geocentric universe and had made all of those painstaking measurements in order to provide support for that view. Ironically, in Kepler's hands, Brahe's measurements became the clinching argument for the Copernican system.

The motion of Mars had been measured in detail by Brahe. There seemed to be no combination of circular orbits of Earth and Mars that gave a reasonable fit to the data. Kepler therefore dropped the notion of uniform circular motion and assumed (1) that the planets move in elliptical orbits with the Sun located at one focus of the ellipse. This assumption was found to be consistent with Brahe's data if (2) the speed of the planet in its orbit was proportional to the reciprocal of its distance from the Sun, which can be shown to imply that equal area segments of the ellipse are swept out in equal times. Finally, Kepler found that (3) the square of the period of such orbital motion is proportional to the cube of the semi-major axis of the elliptical orbit. It was Kepler's empirical laws of planetary motion and Galileo's and Descartes' work in mechanics that provided the inspiration for Newton to formulate the general principles of mechanics and the law of gravitation. Kepler spent his last years pondering the implications of his laws. He arrived at a qualitative notion of an attractive force between two bodies, but it was left to Newton to take the final step.

Sir Isaac Newton (1642-1727) was a moody temperamental man with a morbid fear of controversy and confrontation. He was a fanatic mystic and a great rational scientist, all rolled into one complex person. He never knew his father, who died two months before he was born. His mother's family, who were well-to-do, sent him off to Cambridge to study for the clergy.

Cambridge was one of the world's great universities in the days before Newton, even as it is today. In Newton's time, however, Cambridge had fallen into the intellectual pits. The Aristotelian canon still ruled. In Newton's third year, he discovered mathematics through Isaac Barrow, the new Lucasian Professor of Mathematical Philosophy. Within six months, Newton had mastered all of the mathematics then known and, a few months later, invented *fluxions*, the calculus. Six years later, Barrow passed the Lucasian chair along to a 26 year old Newton. Stephen Hawking is the present holder of that chair as I write this.

Newton worked in isolation, shunning all human contact. For 20 years, only bits and snatches in his correspondence hinted at what he was up to, but that little bit was enough to give him the reputation of a genius.

The seat of English science was in London at the Royal Society. Robert Hooke (physicist), Edmund Halley (astronomer), Christopher Wren (architect), and John Locke (philosopher) were all members of this circle. Unlike Cambridge, which was steeped in mysticism, the Royal Society was a secular institution whose members mostly believed in God, but not in a God who meddles ongoingly. They admired the mechanical philosophy of Descartes, but distrusted his attachment to the teachings of Aristotle. They saw in Newton a champion to meet the reactionary threat.

Newton was a most unlikely champion, singular and moody as he was. By the age of 40, he had given up physics and mathematics and had immersed himself in alchemy and Biblical prophecy. It was Edmund Halley who persuaded him to publish his work, carried out 20 years before. In 1686, the *Philosophiae Naturalis Principia Mathematica* appeared.

In this great book, among many other things, Newton solved a problem for which Christopher Wren had offered a prize. The problem was to explain, and not merely describe, the motion of the planets. Newton had found the key to this while still in his 20s, but now he had the maturity to generalize the result.

The works of Galileo, Descartes, and Huygens had all been missing a key element - the *cause* of motion, the *force*. Newton stated his three laws of motion as:

The momentum (mv) of a body is unchanged unless a force acts upon it.
The force is equal to the time rate of change of the momentum.
For every action (change of momentum), there is an equal and opposite reaction.

The first of these laws is Galileo's Principle of Inertia. Newton only realized this in general during the writing of the final draft of the *Principia*. The third law is Newton's statement of Descartes' (more likely, Huygens') momentum conservation law. The second law, however, was pure Newton, and it is this law that ties the other two into a complete dynamical system.

As if this were not accomplishment enough, Newton also spelled out the law of gravity in the *Principia*.

Galileo had shown that bodies fall with constant acceleration, and he had measured this acceleration. It is about 32 feet per second squared (9.8 meters per second squared) at the surface of the Earth. The acceleration is the same for all bodies, so the force of gravity must be proportional to the mass. By virtue of Newton's third law, the force between two bodies must therefore be proportional to the product of their masses. Newton was further able to show that in order for his law of gravitation to be consistent with Kepler's third law, it was required that the force diminish as one over the square of the distance between the centers of two spherical masses.

As impressive as this sounds, Halley and Wren had both arrived at the same conclusion. Anything that radiates equally in all directions will go as one over the square of the distance. What was needed was to *derive* all three of Kepler's empirical laws, and only Newton could do that.

The accomplishment that really made Newton known to all levels of society was finding the connection between Earthly motions, such as the fall of an apple from a tree, and heavenly motions, such as that of the moon in its orbit about the Earth.

Newton knew that the moon circled at a distance of 60 Earth radii from the center of the Earth, and that it moved at 1016 meters per second along its orbit. He was therefore able to calculate its acceleration, which he found to be 0.00272 meters per second squared. If the one over distance squared law is universally applicable, then the moon's acceleration toward the Earth should be 60 squared or 3600 times less than the acceleration of a falling apple. We easily see that 9.8 meters per second squared divided by 3600 is just 0.00272 meters per second squared, which agrees with the independently determined acceleration of the moon. Thus Newton showed that his law of gravitation was truly universal.

There were a few problems with his mechanics of which Newton was aware but to which, he did not have answers. He knew that steady *velocity* is always *relative*, but *acceleration* appears to be absolute. If you are riding as a passenger at a steady speed in a quiet automobile over a smooth road, and your eyes are shut, you have no way to estimate the speed. If, on the other hand, the car is accelerating, then you feel it immediately. Newton's laws apply only in *inertial* (unaccelerated) frames of reference, and must be patched up by including 'fictitious' forces (inertial forces such as the centrifugal or Coriolis forces) in a non-inertial frame. This disturbed Newton, and he hand-waved the problem away. Bishop Berkeley in the eighteenth century and Ernst Mach in the nineteenth century both considered this problem, to which we will return in Chapter Five.

The other worrisome point was the question of how gravity links the Earth and Moon across (presumably) empty space. Newton assumed that there must be some transporting agency, but he had no idea what it might be. This phenomenon came to be referred to as 'action at a distance'.

Most of Newton's contributions to physics were made before the turn of the eighteenth century. From that time until his death he dabbled in philosophy and mysticism, reverting from Newton, the scientist, to Newton, the magician.

Newton died on the 20th of March, 1727 at the ripe old age of 85. He would have lived longer had he not poisoned himself with mercury in the course of conducting alchemical experiments.

Further progress in mechanics in the eighteenth century consisted of alternate formulations of Newton's laws of motion. Lagrange (1736-1813) presented a second-order differential equation that lent itself to a certain class of constrained dynamical problems. Euler (1707-1783), together with Bernoulli (1700-1782), derived the law of conservation of angular momentum. Euler also did pioneering work in the calculus of variations, while Bernoulli is best remembered for the law of hydrodynamics that bears his name and for work in kinetic theory.

In the field of heat, Fahrenheit (1686-1736) and Celsius (1701-1744) proposed their respective temperature scales, and Joseph Black (1728-1799) measured the heats of vaporization and fusion for water, thus founding the science of calorimetry. Theoretically, physics actually lost ground during the eighteenth century. Newton had correctly regarded heat as being associated with the motion of atoms, however the caloric theory of heat as a subtle fluid of some sort dominated the 1700s.

The nineteenth century was very fruitful for physics and nearly all known phenomena were satisfactorily described theoretically by 1900.

Hamilton (1805-1865) developed a formulation of Newton's mechanics featuring a pair of first-order equations and established momentum as the basic variable in place of velocity. The hydrodynamics of inviscid fluids was developed during this period, although many of the refinements in this difficult field (the equations are nonlinear) awaited the impetus provided by the invention of the airplane.

The conservation of energy is a fundamental concept in physics, however it was not seriously proposed until around 1850. The reason for the delay was that energy can assume many forms. Although energy conservation was known for elastic mechanics, the caloric theory had so stunted progress in heat and thermodynamics that the equivalence of heat and mechanical work was not recognized until the mid-nineteenth century.

The first qualitative experiments on the nature of heat were performed in 1798 by Count Rumford (1753-1814), an expatriate American who emigrated to England and thence to Bavaria where he served as a military engineer. He noted that large amounts of heat were generated in the boring of cannons, too much to be accounted for by the caloric theory. Rumford also noted that the weight of the finished cannon plus the chips from the boring equaled the weight of the cannon before boring. The specific heat of these chips and that of the original metal was the same. He therefore concluded that heat is not a material substance.

A few years later, Sir Humphery Davy (1778-1829) found that two pieces of ice rubbed together under vacuum melted even though the surrounding walls of the vacuum chamber were maintained at a temperature less than zero degrees

Celsius. There was no way in which the caloric fluid could have entered the ice. The caloric theory was surprisingly robust, however. Even Carnot (1796-1832), in proposing his ideal cycle for heat engines in 1824, based it quite unnecessarily on the caloric theory. A critical quantitative experiment was needed.

Joule (1818-1889) provided it. He carried out a careful series of experiments in England wherein the mechanical energy of a falling mass was efficiently converted into a measurable amount of heat by a paddle wheel rotating in water. He found that 778 foot-pounds of energy was equivalent to the amount of heat required to raise the temperature of one pound of water by one degree Fahrenheit. This result was presented as a paper at the meeting of the British Association for the Advancement of Science in 1847. Lord Kelvin, alone of those assembled, recognized the implications and his support was largely responsible for the rapid acceptance of Joule's work. Working independently, Helmholtz (1821-1894) read a paper at Berlin (also in 1847) before the Physical Society in which he proposed the general conservation of energy. The paper was rejected for publication by the prestigious journal, *Annalen der Physik*. Within the next few years, however, the mechanical theory of heat triumphed. Progress now became rapid. The second law of thermodynamics was written in 1850 by Clausius (1822-1888), and in 1851 in another form by Kelvin (1824-1907).

As the nineteenth century drew to a close, the physical understanding of matter and its interactions seemed to be complete and flawless.

CHAPTER THREE

THE GHOST IN THE MACHINE

in which we look at the classical physics
of 'not stuff'.

Nonmaterial manifestations are, by definition, not nearly so demonstrative as material manifestations. In light of this, it is remarkable that the understanding of magnetism, electricity, and light was seemingly complete by the end of the nineteenth century.

In the field of optics, Newton discovered the light spectrum (dispersion) in 1666. He is also known as the inventor of the Newtonian telescope, a form of reflecting telescope from which today's large astronomical instruments are descended. Newton's famous book on *Opticks* was published in 1704. Newton favored a corpuscular theory of light, unfortunately for Hooke (1635-1703) and Huygens who were trying to introduce their wave theories at the same time.

Bradley (1693-1762) made the first reasonably accurate measurement of the velocity of light. As in the case of heat, the eighteenth century was theoretically barren as far as light was concerned. Newton's attachment to the corpuscular theory spoiled the climate for original ideas in this field. Neither were any significant experiments carried out during this century.

Electrostatics, however, was a busy field. Du Fay (1698-1739) showed that flames conduct electricity. The electroscope and the Leyden jar were invented at this time. In the last half of the eighteenth century, three men dominated the work on electricity.

Benjamin Franklin (1706-1790), a versatile American, began his scientific work around 1745. He noted that sharp points or edges of highly charged bodies "throw off electrical fire." Franklin developed a fluid theory of electricity similar in philosophy to the caloric theory of heat; a body with too much of this fluid was considered to be positively charged and a deficiency of it resulted in a negatively charged condition. The famous kite and key experiment was performed around 1755.

Henry Cavendish (1731-1810) contributed to electrostatics and chemistry. He is also known for the Cavendish experiment to measure G, the gravitational constant, first carried out in 1798. Cavendish was a remarkably productive scientist, but he kept most of his discoveries to himself. The extent of his achievements was not appreciated until his notes and manuscripts were finally sorted out and published by Maxwell in 1879. Cavendish found the law relating the force between two charges (independently of Coulomb), conceived the capacitor, was acquainted with the notion of electrical potential, and wrote Ohm's law 50 years before Ohm.

As the nineteenth century dawned, a strong revival of the wave theory of light was begun by Thomas Young (1773-1829) in his work on interference. Unfortunately, the scientific establishment was not yet ready for light waves. Young was attacked and ridiculed by his contemporaries. In 1818, Fresnel (1788-1827) rediscovered the phenomenon of interference. He suggested that the light vibrations in the assumed aether transmission medium (about which we will have more to say in the next chapter) were transverse rather than longitudinal with respect to the wave vector. This was soon verified when polarization phenomena were discovered. In 1850, Foucault (1819-1868) demonstrated experimentally that the velocity of light in water is less than the vacuum value as required by the wave theory. This pretty well clinched the argument and vindicated Young.

Electromagnetism was an especially fruitful field of endeavor in the nineteenth century. Galvani (1737-1798) had carried out his famous frog leg experiment (the excitation of muscular contraction by an electric current passed through the leg of a deceased frog) in 1786. He also discovered the so-called Galvanic current between dissimilar metals. Volta (1745-1827) made the first wet cell in 1800 using zinc and copper plates separated by blotting paper soaked in brine. He soon found that such cells could be wired in parallel and series batteries to provide a source of considerable electrical power. This was an invaluable contribution to electromagnetic research since magnetic fields require a current source that had not been available prior to Volta's discovery. Volta himself, no doubt following the lead of Galvani and his frog leg, constructed a battery of considerable voltage and placed a wire from it into either ear. He reported a vivid flash and a noise "...like thickly boiling soup..."

With a current source now available, the arc and the Ohmic heating that takes place in a current-carrying resistor were soon discovered. The Ampere and Biot-Savart laws relating the current and the magnetic field were shortly found and Ampere (1775-1836) discovered the electromagnetic force law that makes electric motors possible. Michael Faraday (1791-1867), the son of a blacksmith, built the first electric motor in the early 1820s. Faraday's best known achievements were in the area of electrical induction. Oersted (1777-1851) had shown earlier that electricity (current) can produce magnetism. Faraday believed that the converse was also true. His search for a steady current in the scheme shown in Fig. 3-1(b) failed. In 1824, Arago (1786-1853) discovered the drag due to eddy currents when a conducting disc is moved into or out of a magnetic field. Shortly thereafter, Faraday discovered that when the primary coil of the apparatus shown in Fig. 3-2 was energized or switched off, the needle of a compass placed close to the secondary coil circuit was momentarily deflected, indicating that a current existed in the secondary only for the brief time when the switch was opened or closed. Faraday made the association with the Arago experiment and also found that the effect could be produced by inserting or withdrawing the bar magnet from the coil of Fig. 3-1(b). The conclusion was that it is not the magnetic field that induces the current, but the *change* in the magnetic field that does it.

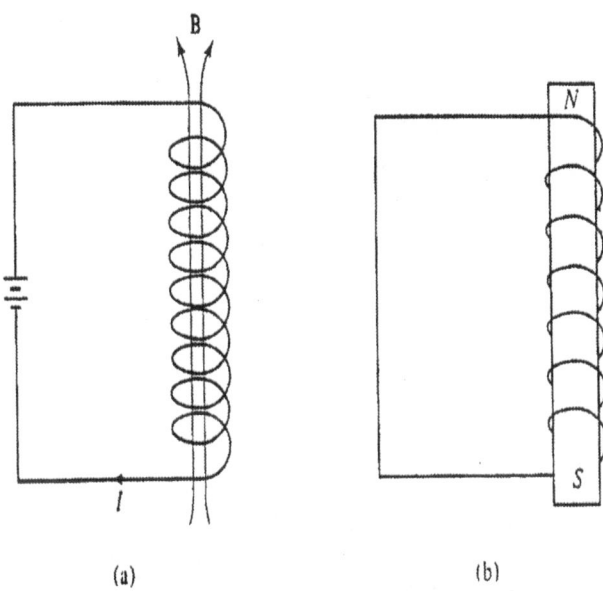

(a)   (b)

Figure 3-1 (a) When a current (I) is passed through a coil, a magnetic field (B) is established in the coil. (b) Faraday reasoned that magnetic field established in a coil, albeit by a bar magnet, might result in the induction of a current in the circuit. Nature refused to put its stamp of approval on this notion.

Faraday's knowledge of mathematics was minimal. His successes were based largely on a very keen physical intuition. It was he who first conceived the notion of field lines and "tubes of force". The abstract idea of fields is now central to theoretical physics, and it was conceived by an experimentalist with no formal education.

Joseph Henry (1799-1878), a contemporary of Faraday's, was an American who taught mathematics at the Albany Academy. His scientific work was performed largely during the one summer month each year when his teaching duties allowed. America at that time had almost no scientific activity (Europe was still the seat of learning), and thus Henry lacked the stimulation that association with other scientists would have given him. Nevertheless, he managed to construct an electric motor and to discover the laws of induction quite on his own. It was the work of Joseph Henry on electromagnets that led Samuel F. B. Morse (1791-1872) to invent the telegraph.

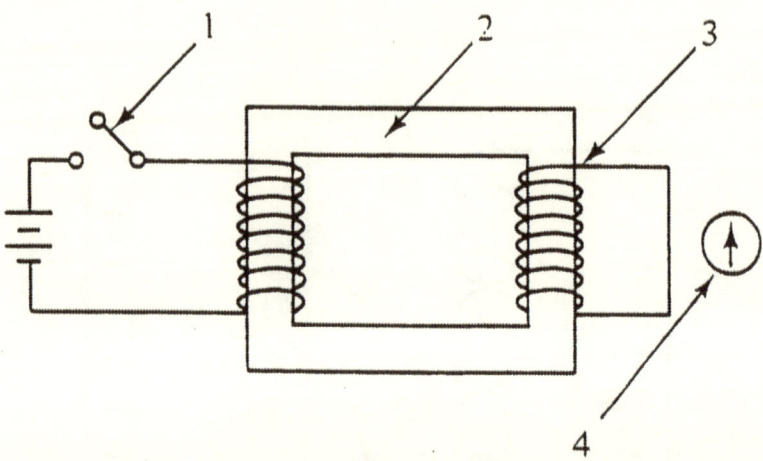

Figure 3-2 Faraday induction apparatus. When the switch (1) in the primary circuit is either opened or closed, the change in the magnetic flux through the iron core (2) induces a current in the secondary winding (3) that momentarily causes the compass (4) to deflect.

Almost all of the pieces of the electromagnetic puzzle were there and it was James Clerk Maxwell (1831-1879) who supplied the final piece and pulled it all together. Maxwell was born into a wealthy Scottish family in Edinburgh. He graduated summa cum laude from Trinity College at Cambridge in 1854 and had already published a number of papers by the time he was 19 years old. From 1860 to 1865, Maxwell held a professorship at King's College, London. It was during this time that he published his *Dynamical Theory of the Electromagnetic Field.* In 1870, he moved to Cambridge as professor of experimental physics. In this position, he established the Cavendish Laboratory and served as its director until his untimely death at the age of 48. Maxwell did pioneering work in a number of fields. He was interested in the means by which the eye perceives color. His work in kinetic theory, together with that of Clausius, resulted in the comprehensive theory we use today.

It is primarily Maxwell's contribution to the field of electromagnetism that will concern us. The key piece to the puzzle was supplied with Maxwell's proposal of the displacement current. Just as Faraday had shown that changing magnetic fields give rise to electric fields, so Maxwell had been able to demonstrate that changing electric fields give rise to magnetic fields. The latter effect is smaller than the former by a factor of one over c squared, where c is the speed of light. It is just this symmetry between electric and magnetic fields that gives rise to the feedback necessary for the existence of electromagnetic waves. Maxwell found that these waves travel at the speed of light and showed that all the phenomena of optics (reflection, refraction, etc) could be derived from the electromagnetic equations. It was now obvious that visible light waves were just one part of the electromagnetic spectrum. Maxwell published these results in 1864. Heinrich Hertz (1857-1894) produced the first non-optical electromagnetic waves in 1887, using a spark gap and an induction coil. In 1895, Marconi (1874-1937) produced the first wireless telegraphy system. On December 12, 1901, he successfully transmitted the first transatlantic radio signal.

And so we find at the close of the nineteenth century that physics had reached a pinnacle of success. Newton's mechanics was unchallenged and gave accurate answers to all problems from the kinetic theory of colliding molecules to the orbital motion of the planets. The wave theory of light had triumphed over the corpuscular theory and had been integrated into the general electromagnetic theory. Except for a few trifling details (some uneasiness about applying Galilean transformations to the electromagnetic equations and difficulties in working out a theory for thermal radiation), everything seemed to have fallen neatly into place. It seemed clear that we live in a deterministic, objective universe where, if we had a complete knowledge of any system at one particular time, it would, in principle, be possible to calculate the details of the system at any other time in the past or the future. This satisfying clockwork machine we call the Universe was played out on a fixed stage of universal space and time in front of us, an audience of independent observers.

This is all well and good so long as we do not apply this same view to ourselves, for then we must see ourselves as robots whose every thought and action are totally predetermined, beings with no free will at all. Clearly this is unacceptable on every imaginable level, and so we agreed, with no really workable argument, to except ourselves from the view that we applied to the rest of the world. What we really did, of course, is to agree not to think about it.

Having put that little bit of unpleasantness behind us, it is fair to say that in the year 1900, most people felt that physics' work was done. Clearly, the twentieth century would belong to the engineers. They could hardly have been more wrong.

CHAPTER FOUR

CRACKS IN THE FOUNDATION

in which we pay the price for
consistency between our theories of
'stuff' and 'not stuff'.

At the end of the nineteenth century, there were two great theories in place
that, between them, appeared to explain all physical phenomena. There was,
however, a small question of mutual compatibility. To illustrate the problem,
let's start out by looking at how sound waves move through air.

First, let me remind you that Newtonian mechanics applies in any
unaccelerated frame of reference, that is, in any frame that moves in a straight
line at constant speed. We noted that such a frame of reference is said to be
inertial. You can play ping pong in a room of your house or in an airplane flying
steadily at 600 miles per hour and the results are the same. There does not appear
to be a preferred inertial frame. One would seem to be quite like another so far as
any mechanical experiment can reveal.

Waves, as described by Newtonian mechanics, are disturbances that move in
some medium at a given speed. With the benefit of this information, imagine
now that we are in a moving train and a loudspeaker is set up at one end of the
enclosed train car. We measure the speed of sound waves generated by the
speaker and find something on the order of 1100 feet per second (335 meters per
second). Although the train is moving, the air, the speaker, and the experimenter
in it are all moving along together, so we obtain the result that corresponds to the
known speed of sound in still air.

Suppose now that the loudspeaker was located beside the tracks and we were
riding in the train moving at, say, 400 feet per second in the direction of
propagation of the sound wave. What speed would we measure for the sound
wave now? The wavefront is moving at 1100 feet per second and we are chasing
it at a speed of 400 feet per second, so the speed with which the wave is pulling
away from us is just going to be 1100 minus 400 or 700 feet per second. No
mystery there.

Now let's go from sound waves moving through air to electromagnetic
waves (or light) moving through whatever it is that they move through.
Maxwell's theory tells us that such waves move at a speed of 186,000 miles per
second (about 300,000 kilometers per second). On the basis of our experience
with sound waves, we are justified in asking two questions. First, in what

medium are the electromagnetic waves propagating (what is doing the waving?), and second, with respect to what frame of reference is the speed of propagation 186,000 miles per second? Maxwell recognized the validity of these questions about which the electromagnetic theory itself was silent. In an article in the *Encyclopaedia Britannica* he postulated a medium, which he called the aether, that pervades all the space between the stars and between the atoms and serves as the carrier of the light waves. The frame of reference in which the speed of light has the value predicted by Maxwell's theory was just the frame in which the aether is at rest. On the basis of this reasoning and our discussion of sound waves, it should be possible to record a variation in the measured speed of light as the Earth moves in its orbit about the sun. Even if the Earth were moving with the aether in June, it would certainly be moving against it in December.

What sort of properties must the aether have? In order to respond to fast high-frequency waves, a medium must be very stiff. For example, the speed of sound in vulcanized rubber (which is not at all stiff) is only 177 feet per second (54 meters per second), whereas in granite, sound can move at almost 20,000 feet per second (about 6100 meters per second). The aether, in propagating light waves at 186,000 miles per second, must be much stiffer than granite. At the same time, we know that the Earth and the other planets have been moving about the sun in their orbits for billions of years with no apparent drag, so no matter how stiff the aether is, it must be almost perfectly inviscid in order that the heavenly bodies not be slowed in their orbits. It is hard to imagine that any substance could have such bizarre properties, but it would seem that there must be an all-pervading aether to carry the electromagnetic waves and it must have such properties.

An American physicist, A. A. Michelson (1852-1931), carried out an experiment in 1881 to detect the aether while on sabbatical in Germany from the U. S. Naval Academy. Michelson set out to measure the stream speed of the aether 'river' with respect to the Earth. For this purpose, he developed an optical device known as an interferometer. This is shown schematically in Fig. 4-1. The Michelson interferometer manages, with the use of a half-silvered mirror, to split a light beam into two beams of approximately equal intensity. These beams are sent over two equal paths at right angles to each other and then recombined on a ground glass screen.

There is something that you should know about electromagnetic waves that is also true of water waves, with which we have more direct experience. You can simply add their amplitudes. Physicists say that such linear waves are superposable, which means the same thing. Where a wave peak coincides with another wave peak, they add to produce a higher peak. Where a wave trough coincides with the peak of another wave, the two waves cancel and the water stays momentarily flat. This phenomenon is referred to as interference. If the times required for the light beams to travel the two paths in the same, this would reveal itself at the point of observation as an interference fringe shift, a slight displacement of dark bands, which are caused by the mutual interference of the light waves.

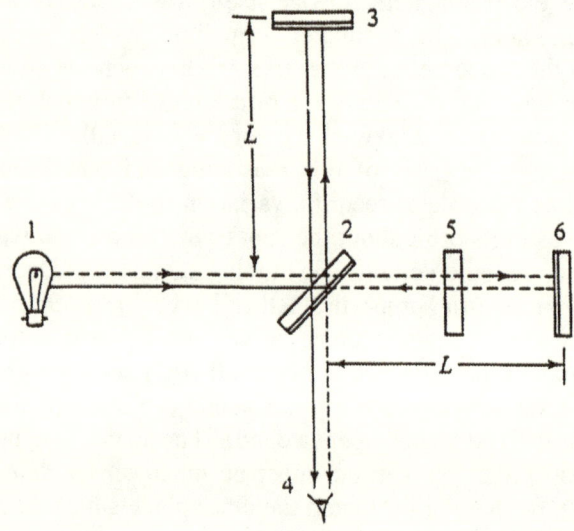

Figure 4-1 The Michelson interferometer. Light from a monochromatic source (1) is spilt into two beams by a 45-deg mirror (2), which is lightly slivered on the back side such that the two emergent beams are at right angles and have appoximately equal intensities. The cross-stream beam is reflected from a front-silvered mirror (3) located a distance L from the beam splitter. The beam returns to the beam splitter and half of it passes through to arrive at the observation point (4). The upstream-downstream beam passes through a glass plate (5), which is there to ensure equality of optical path for the two beams, and thence to a front-silvered mirror (6) where it is reflected back to the beam splitter. Half of this beam is then reflected to the observation point where it is combined with the cross-stream beam. Any difference in the time of flight of the two beams is evident by studing the interference pattern.

In Michelson's 1881 experiment in Berlin, the anticipated fringe shift owing to the motion of light with respect to the aether was about 0.04, which was on the same order as the experimental error, so no conclusions could be drawn from the fact that no fringe shift at all was seen.

When Michelson returned to America, he did not go back to Annapolis, but instead accepted an appointment at the Case School of Applied Science in Cleveland, Ohio. In 1887, He repeated his earlier experiment, working in collaboration with Edward W. Morley. In order to produce a definitive result, the optical path length was increased from the 1.2 meters of the 1881 experiment to 11 meters by the use of multiple reflections. This gave an anticipated fringe shift of 0.4, which was some 20 to 40 times larger than the experimental error. Again, however, the measured fringe shift was zero. The experiment was repeated at various times of the year, always with a null result. These findings were taken quite seriously by the physics community and the mystery deepened. The postulated aether stream refused to reveal itself.

A number of more or less desperate explanations were offered for the failure of the Michelson-Morley experiment to detect the aether. In 1889, Irish physicist George F. Fitzgerald proposed an explanation. He said that the leg of the experiment in the direction of the Earth's motion contracted by just the amount needed to give the null result. In effect, Fitzgerald was suggesting without proof that the electric force that binds matter together was slightly stronger in the direction of the motion and that this stronger component shrinks the experiment (and the experimenter).

This notion received support from Hendrik A. Lorentz in the decade following the 1887 experiment. Lorentz showed that the electromagnetic fields, when viewed by an observer moving at a uniform velocity with respect to the aether frame, were altered in just such a manner as to produce the contraction proposed by Fitzgerald. This put the idea on a much better footing since Lorentz's theory was not merely an ad hoc explanation of the Michelson-Morley result. Both Fitzgerald and Lorentz continued to believe that there must be an aether and to believe that the Earth moved through it with a finite velocity, even though such motion was inherently undetectable.

Henri Poincaré, carried Lorentz's ideas one step further. He agreed that absolute uniform motion was undetectable. He also noted that it must be possible to write an entirely new dynamics in which the speed of light would appear as a limiting speed.

Poincaré, was on the right track, as we shall see, but he did not develop these ideas. It remained for a minor clerk in the Swiss Patent Office to take the critical step.

In 1905, Albert Einstein (1879-1955), a recent graduate of the Swiss Federal Polytechnic Institute, published a paper (three papers, actually, but more on the other two later) entitled *On the Electrodynamics of Moving Bodies* in which he took the step (giant leap, more like) of dropping the idea of the aether entirely. Nothing is waving, he said. In his paper he set down two principles and proceeded to derive their consequences. These principles were:

Absolute uniform motion is inherently undetectable.
The speed of light is the same in all frames of reference.

In an interview toward the end of his life, Einstein claimed that at the time he wrote his paper he was unaware of the Michelson-Morley experiment, Poincaré,'s work, or Lorentz's theory in its final form. Maybe, maybe not. Einstein's entire scientific career was marked by a strong tendency toward self-sufficiency.

The first of Einstein's two postulates looks reasonable. Fitzgerald, Lorentz, and Poincaré, would all have agreed with it. The second postulate, which really just states the implications of the Michelson-Morley experiment, has immediately evident consequences that defy our reason.

This is a good place to make a point. The reason that the second postulate disturbs us is that it runs counter to our common sense based on gut experience in the world of appearances. Bear in mind, however, the admonition of Sir Oliver Lodge who warned against trusting common sense too far when he remarked that our senses were developed by the struggle for existence, not for the purpose of philosophizing on the world.

Suppose now we go back to the example previously given of the train. This time we assume that there is a lamp beside the track that is emitting a beam of light in the direction of motion of the train. The light moves at a speed c equal to 186,000 miles per second with respect to the Earth. Suppose that the train is moving at a speed, say, of 0.9c. What speed will an observer on the train measure for the speed of the light beam? Will he measure c minus 0.9c equal to 0.1c as the example with the loudspeaker would suggest? No, according to Einstein's second postulate, all observers moving uniformly in whatever direction and at whatever speed must measure 186,000 miles per second as predicted by Maxwell's theory. According to Einstein, this result implies that the aether is an invalid concept.

Einstein's postulates have bizarre consequences for our notion of space and time. Let's see what they are. We have been duly warned not to trust our intuition, so we are going to have to dot all i's and cross all t's very carefully. The first thing we need is a very simple and trustworthy clock. We will take advantage of Einstein's second postulate and use a beam of light, the propagation speed of which all observers agree is c. Figure 4-2 shows this clock. It consists of a light source mounted in one end of a box, the other end of which is fitted with a light detector of some sort. One tick of the clock is just the time that it takes light to travel the length of the box.

Suppose now that the clock is set into motion to the right as shown in Fig. 4-3. What period of time do we measure for a tick now? Bear in mind that the light now has to travel not only the length of the box, but also the sideways distance moved by the box while the light is in flight. Thus a moving clock runs more slowly and as the speed of motion of the clock approaches the speed of light, time appears to stop altogether.

A few more words about this *time dilation* effect are in order. The time period between events occurring at the same place in a reference frame as measured by a clock at rest with respect to the frame is called *proper time*. In other words, the wristwatch on your arm keeps proper time for you as an observer. All time periods measured in other frames of reference moving with respect to the observer are longer; that is, moving clocks are observed to run more slowly. This effect applies to all types of clocks (electronic, mechanical, and biological) and is a direct result of the finite and invariant speed of light.

Now let us look at the spatial consequences of Einstein's postulates. Consider a rod moving in a direction parallel to its length at some speed v. An observer moving with the rod would measure its proper length (proper in the sense just defined) using a meterstick in the conventional manner. A stationary observer who watches the rod move by, starts his stopwatch when the leading end of the rod moves past his position and stops his watch when the trailing end goes past. The time recorded is the proper time for the rod to pass. The stationary observer knows the speed of the rod to be v (by use of Doppler radar, perhaps) and from his measurement of the time taken for the rod to pass he can calculate the length L, which is just the product of the speed v and the measured proper time. Using the meterstick and a clock moving with the rod, the stationary observer now finds the proper length as the product of the speed v and the dilated time. Since v is the same in both cases, the implication can be seen to be that moving lengths contract by the same factor that moving clocks slow down. This is, of course, a consequence of the fact that the speed of light must be the same in all frames of reference. The contraction is just that proposed by Fitzgerald, only now we have an explanation of it.

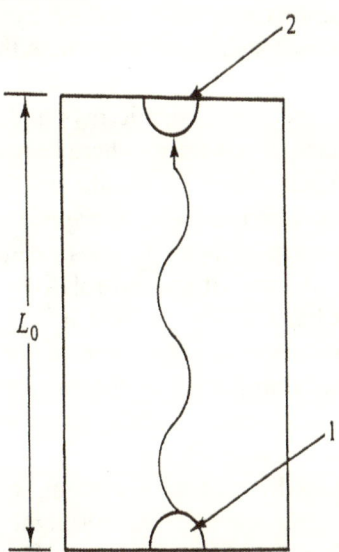

Figure 4-2 A light clock. Light emitted by a source (1) travles the length of the box L_0 in a time $T_0 = L_0/c$ and is detected by a photomultiplier (2). The time T_0 is one tick of the clock.

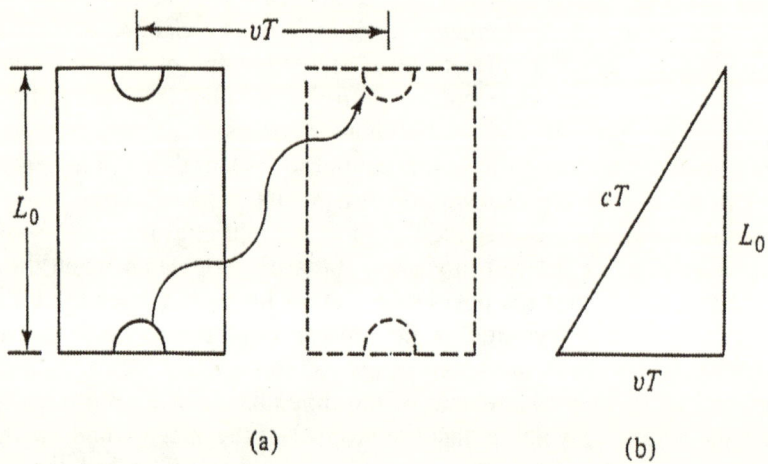

(a)   (b)

Figure 4-3 The light clock of Fig. 4-2 moving to the right at speed v with respect to the observer.

Contracting rods and clocks that vary their rates are not easy concepts to swallow, but these predictions are well verified by experiment. Consider the following example. A muon is a sub-atomic particle that, when created at rest, is known to have a certain lifetime until it decays. Muons are created at the top of the Earth's atmosphere by incoming cosmic rays and the fast muons manage to make their way to the surface of the Earth where they are detected. If we were to think about this in Newtonian terms, we would say that the distance traveled is such that even if the muons were moving at the speed of light, they could not live long enough to make the trip. Their lifetimes seem to have somehow been extended. From the point of view of special relativity, the lifetime clock of the rapidly moving muon has been slowed so that it has the time required to make the journey. From the point of view of the muon (if we suppose that a muon has a point of view), the distance from the top of the atmosphere to the Earth's surface is a moving length and it is contracted sufficiently to permit the journey to be completed within the normal resting lifetime.

As a further example to drive home the point, suppose that you wanted to travel within your lifetime to a star that was 500 light years away (a light year is the distance traveled by light in one year). From a Newtonian point of view you would say, no, it can't be done. Even if I traveled at the speed of light it would take 500 years to complete the journey. Suppose you had at your disposal a space ship capable of a speed of 0.999c. According to the equations of relativity you could make the trip in a time (as measured by you) of 22 years and four months because, for you, the distance of 500 light years would have shrunk. To the people you left back on Earth, 500 years and six months would have gone by and you and your voyage would probably have been long since forgotten. In this way, special relativity is a time machine that allows you to travel into the future. All of these wonders are consequences of exchanging an inflexible space and time and a variable speed of light for a flexible space and time and an invariable speed of light.

Now let's examine another aspect of relativity and see what it has to say about simultaneous events. Simultaneous events are, of course, events that happen at the same time and we would think that all observers would agree that this is so. Einstein says no. Suppose, for example, that we find ourselves in a large glass box moving sideways at high speed. In the exact center of the box is a light bulb that flashes on every few seconds. Since the front (leading in terms of the motion) and rear walls are equidistant from the light bulb, we would say that the light reaches the front wall at exactly the same time that it reaches the rear wall. In other words, those two encounters are simultaneous events. Now let us take the point of view of an outside observer who sees the glass box rushing past and suppose that the bulb flashes just as the box is opposite his position. He will see the front wall moving away from the oncoming light and the rear wall moving toward it. For him, the light will reach the rear wall before it reaches the front wall and he would claim that the two events are not simultaneous.

This revelation is more jarring than you might suppose. We all know that time has a past that exists only in our memories and a future that has not yet occurred. The interface between past and future is the present, which has no duration at all. Unlike space, in which we can go forward or backward, time is a one-way street and proceeds (in our experience) only from past, through present, to future. For the man in the glass box, the light strikes both walls simultaneously in his present time. The stationary observer sees the light strike the rear wall in his present time. The event of the light hitting the front wall is still in his future. For him, when the light hits the front wall, the encounter with the rear wall is already a past event stored in his memory. The lesson is that the relativity of simultaneity breaks down our notion of a universal past, present, and future on which all observers agree.

Special relativity is counterintuitive, to say the least. Thus it is very helpful to be able to picture some of the relativistic effects, such as the breakdown of simultaneity. This can be done using a *Minkowski diagram*. Hermann Minkowski was one of Einstein's teachers at the Swiss Federal Polytechnic Institute. In 1908, Minkowski wrote a paper in which he recast relativity in a geometrical form.   Since space and time were no longer universal and independent, Minkowski had the idea to treat time as a sort of fourth dimension in what we now call Minkowski space. He added no new physics, but his geometrization of what came to be known as *space-time* was a useful way to look at special relativity and was an influence on Einstein when he came to write his theory of general relativity.

Now let's get back to the Minkowski diagram that I mentioned. This is shown in Fig. 4-4. What we have are two coordinate axes drawn on top of each other. The time axis is vertical and the axis representing the direction of motion is horizontal. The two spatial axes perpendicular to the motion are ignored. The x, t axes describe the system at rest and the x', t' axes describe the moving frame. The scale is such (c = 1) that the light line, x = ct, is a 45-degree line. Note that whereas the x and t axes are mutually perpendicular, the x' and t' axes are not. The figure shows two events, A and B, that occur simultaneously in the moving system, that is, they both lie on the x'-axis, which is t' = 0. When we project these events into the stationary frame by drawing dashed lines from them parallel to the x-axis (t = 0), we see that they intersect the t-axis at two different times separated by Δt. This gives a picture of the glass box thought experiment that we discussed.

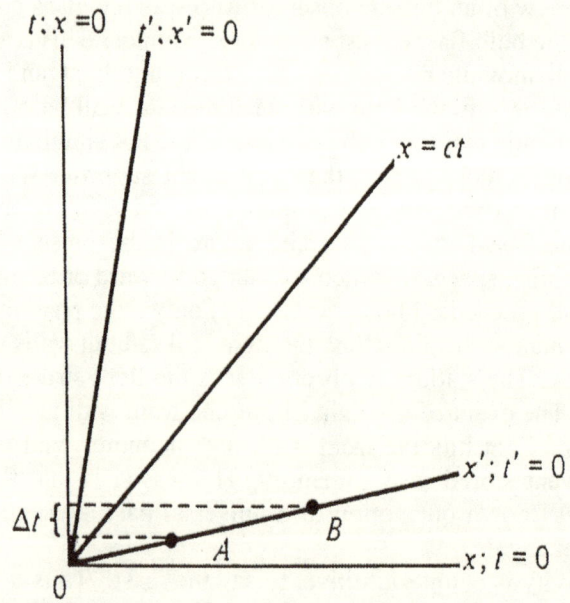

Figure 4-4 The Minkowski diagram showing the breakdown of simultaneity.

The next thing that I would like to talk about is the effect of relativity on energy and momentum. First, let me define momentum for you and show why it is such an important variable in mechanics.

We saw in Chapter Two that the second of Newton's laws of motion states that the time rate of change of the momentum, p, of a massive object is equal to the net force acting upon the object. The momentum is defined as the product of the velocity v and the mass m. For situations where the object, or system of objects, has no net force acting upon it, the momentum will stay constant, or as the physicist puts it, momentum is conserved. If you consider a situation where you have a couple of objects colliding with each other and rebounding in the absence of external forces, the fact that we know that the momentum of the system stays constant enables us to analyze this complicated situation with ease. I said all this back in Chapter Two, but a reminder at this point might be useful. You can see why physicists like to find quantities that are conserved and hate to give them up when a new theory comes along.

In special relativity, the product mv is not conserved. Thus we would like to find something that *is* conserved in relativity that, in the limit of low velocities (the Newtonian limit), goes back to being mv. This can be done and one finds that m times v times the dilation factor that slows moving clocks is conserved relativistically. Force is still the rate of change of momentum and the kinetic energy of an object is still found by taking the product of the force and the distance over which it is applied. When you go through this calculation, you find that there is a residual energy associated with a body at rest, a very large residual energy given by

$$E = mc^2,$$

the one equation of which everyone has heard. It is the most famous equation in twentieth century physics and upon it rests the miracle of nuclear energy. It essentially tells us that the amount of energy potentially available in matter is equal to the mass times the square of the speed of light.

Almost all physics texts claim that the mass of an object increases as the speed of the object increases, approaching infinity as v approaches c. This is not strictly true. As we have seen, our definition of momentum had to be modified by including the dilation factor, but to say that the factor applies only to m and not to the momentum as a whole is a mistake. The mass of a body is a measure of how much of it there is in crude terms and to claim that more stuff accrues as the speed increases is not justifiable and is, in fact, inconsistent with the energy calculation.

The kinetic energy of an object, like its momentum, must approach infinity as the speed of the object approaches the speed of light (owing to the dilation factor). The inference is that massive objects must always move at speeds less than the speed of light. If an object has zero mass, it can only move at the speed of light and at no other speed. Special relativity also allows for the possibility of objects that move at speeds greater than the speed of light. These have been called *tachyons*. If they exist, they must have imaginary mass (imaginary in the sense of the square root of minus one) and they must travel backward in time. Tachyons require an infusion of energy to slow them down and lose energy when they speed up. Most physicists doubt that tachyons really exist and are convinced that even if they did, they would probably be impossible to detect. Maybe, maybe not. We should always bear in mind Murray Gell-Mann's compulsion principle, that states than anything not excluded by physics is compulsory. Let's say that the jury is still out on tachyons.

What about the possibility of propagating information (as opposed to mass) at speeds in excess of the speed of light? This turns out to be a very important question that will come up again in a critical context later on in the book. From an energetic point of view there is no problem since momentum is not involved.

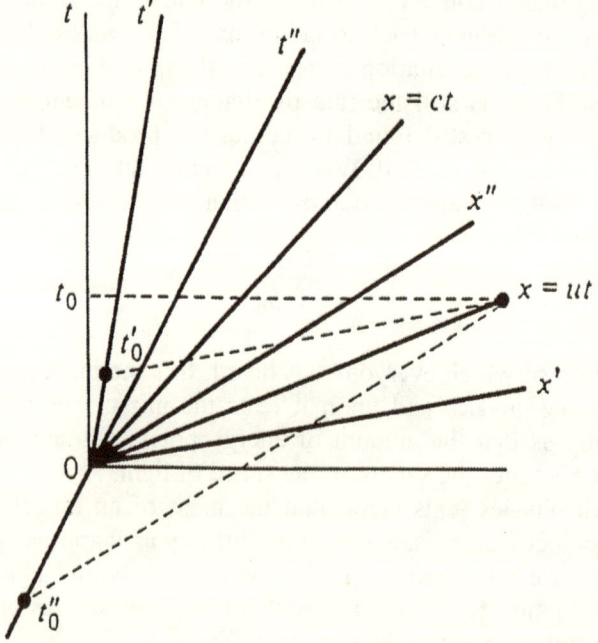

Figure 4-5 The breakdown of causality for superluminal events

In Fig. 4-5, we show a Minkowski diagram that sheds light on this question. There we have superimposed two Minkowski diagrams by including two sets of axes describing moving frames, the primed and the double-primed systems. Suppose that there is an event that is moving in the x-direction at a speed u in excess of the speed of light. We register the event in the stationary (unprimed) system at time t_0. We would like to know at what times (with respect to $t_0 = t_0' = t_0'' = 0$, the time at which all agree that the event was launched) the observers in the moving primed and double-primed systems register the event. To find the answer, we project back parallel to the x' and x" axes respectively and find t_0' and t_0''. The observer in the primed system sees the event at a positive time t_0', but the double-primed observer sees the event at a negative time, that is, before the event is launched. For him, the event has occurred before its cause, and this is something that most people assume cannot happen.

The train of thought goes something like this. Detection of an event before its cause occurred (together with the assumption of free will) implies that one might change the future so as to eliminate the cause of the event that you already detected. In order to preserve self-consistency, most people will be prepared to say that superluminal signals cannot be sent and that one part of the universe cannot know what is going on in another part of the universe in a time less than the time required for light to propagate. This is called the *Principle of Locality*. Keep this discussion in mind. We will have occasion to refer to it later. In fact, it is central to our effort to find out what is really going on under the skin of the world of appearances.

CHAPTER FIVE

THE FUN-HOUSE MIRROR

in which we examine Einstein's General
Theory of Relativity where space-time is
curved and twisted by its contents.

The thing that is special about special relativity is that it, like Newtonian mechanics, applies only in inertial frames of reference, that is, in frames moving at a constant speed in a straight line. What happens when we find ourselves in an accelerated frame of reference? The most familiar example is, of course, straight line acceleration. In addition, we are all familiar with the fact that when we make a change of direction, such as driving our car around a traffic circle, we feel a centrifugal force pulling toward the outside of the curve. From an elementary point of view, this is not a basic force in the same sense as gravity or the electrostatic force between charges. It is an inertial force that acts only so long as we stay on the curve and ceases the moment we straighten up the steering wheel. There is another inertial force with which we are less familiar. It is the Coriolis force that describes motion with respect to a rotating body. To make this clear, imagine that you fire a rocket from one location on the Earth's surface and intend that it land at some predetermined place. While the rocket is aloft, the Earth is spinning beneath it and you will have to take this fact into account by including a fictitious force proportional to the spin of the Earth and the velocity of the rocket. This all sounds mundane, but actually it is rather mysterious.

We found in our discussion of special relativity that there is no absolute frame of reference, but now that we are considering accelerated frames of reference, we have to be open to the possibility that there may be. Consider the following thought experiment. We have two deformable spheres located far enough from each other so that distortions owing to their mutual gravitational interaction are negligible. One of the spheres is caused to rotate about their common axis (see Fig. 5-1). Observers on each spheres would then see the rotation of the other sphere with respect to his own reference frame. Can we say that one of the spheres is rotating in an absolute sense? After our exposure to special relativity, we are tempted to say that the question is meaningless and that the rotation is purely 'relative.' We would be wrong. If the two observers take note of the shapes of their respective spheres, one of them will find that his sphere (A) is still perfectly spherical, while the other sphere (B) has assumed a flattened spheroidal shape. The deformed sphere is rotating with respect to *absolute space* (as Newton termed it), whereas the other sphere is not. So what is absolute space? It is that which gives rise to inertial forces and appears to have no other properties. Clearly, this is not satisfactory. It's like the aether all over again. We need to dig deeper.

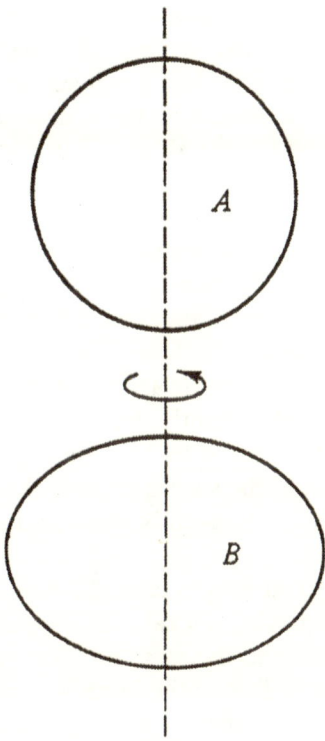

Figure 5-1 Two spheres in relative rotation about a common axis

First, let's get rid of sphere A, the undeformed sphere; it has served its purpose. Sphere B is now rotating in deep space and the equatorial bulge can be measured, so we know that the centrifugal force is at work. What does our observer standing at the pole of the rotating sphere see when he looks up? He sees the rotation of the so-called fixed stars about the extended line of the rotation axis of his sphere just as we do on Earth. They are not really fixed, of course, but they are so far away that their motion relative to each other is indiscernible. So it is the fixed stars with respect to which the sphere is rotating and we can drop this 'absolute space' business. What if the fixed stars were not there and our sphere was alone in an otherwise empty universe? Could there still be said to be a rotation and would the centrifugal force still manifest? This is a tough question, but I think that the answer has to be no. The process of elimination has brought us to this conclusion.

This notion that the centrifugal forces, indeed, all inertial forces, have their origin in the totality of distant masses was first suggested by Bishop Berkeley in Ireland about 20 years after Newton published his *Principia*. The philosopher-physicist Ernst Mach (1838-1916), writing 150 years after Berkeley, made an important extension of Berkeley's idea. Mach suggested that inertial frames are distinguished from noninertial frames by being unaccelerated with respect to the average distribution of matter in the universe. This statement is known as *Mach's Principle*.

We know that mass is both the seat of gravity and that upon which gravity acts, according to Newton's law. Thus mass acts both in an active and a passive way. It is quite the same with electrical charge in Coulomb's law. We also know that mass is that upon which inertial forces act. Mach's principle suggests that the source of the inertial forces is to be found in distant mass. We know that the strength of Newtonian gravity falls off as one over the distance squared so that the effects of gravity are predominantly due to nearby masses. Could it be that gravity, in some way not considered by Newton, also acts so that the effects of distant masses are dominant as the source of inertial effects? This was one of the ideas that was going around in Einstein's mind in the years following his 1905 paper on special relativity.

In 1890, Baron Roland von Eötvös carried out an experiment to measure the relationship between mass playing a local gravitational role and mass playing an inertial role to see if the ratio was the same for a variety of substances. We know that any body at rest on the surface of the Earth is subject to a gravitational force F_g directed toward the center of the Earth, and an inertial (centrifugal) force F_i directed outward at right angles to the Earth's axis of rotation as shown in Fig. 5-2. Eötvös placed two equal masses of different materials on either end of a rod suspended in the middle from a torsion balance. The experiment was located at a latitude roughly midway between the equator and the pole and the rod was oriented east-west. If the ratio of inertial mass to gravitational mass was different for the two materials, then the horizontal components of the two forces would establish a torque that could be measured by the torsion balance. Eötvös observed no such torque and concluded that the ratio of inertial to gravitational mass must be the same for all the materials tested at least to one part in a hundred million. The Eötvös experiment has been repeated many times with the same result and the error is now down to about one part in a trillion, that is, one part in a million million. Thus we can safely conclude that the ratio of inertial mass to gravitational mass is the same for all materials.

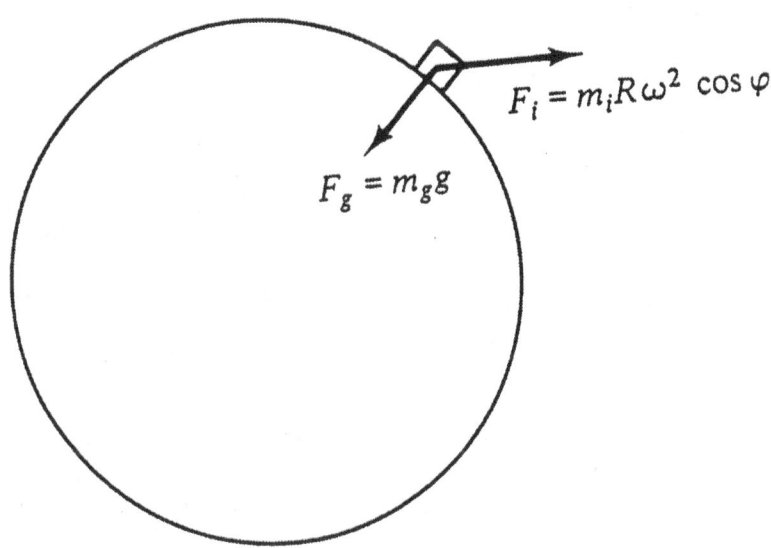

$$F_i = m_i R \omega^2 \cos \varphi$$

$$F_g = m_g g$$

Figure 5-2 The inertial and gravitational forces on an object at rest on the surface of the Earth. The radius of the Earth is R, ω is its angular frequency, and φ is its latitude.

In 1911, Einstein made one of his intuitive leaps on the basis of Mach's principle and the notion that gravity and inertia must be two sides of the same coin. He stated that:

Over a limited region, the effects of being in a uniform gravitational field and of being in a noninertial frame of reference are indistinguishable.

In other words, the pull of gravity feels just like the effects of acceleration. This has come to be known as the *Principle of Equivalence* and though it seems quite similar to Mach's principle, it has far-reaching implications that Mach never envisioned.

First, let's illustrate the principle of equivalence with one of those thought experiments that are so useful. Imagine two elevators (lifts to those of you of the British persuasion), both without windows, each containing a couple of observers and some sort of apparatus for doing mechanical experiments (a ping pong table with paddles and ball will be fine). Let one of the elevators be located in deep space far from any gravitational source. Imagine that its cable is being drawn upward by some cosmic agent in such a way that the elevator experiences an acceleration just equal to the acceleration of gravity at the surface of the Earth, about 32 feet per second squared. Let the other elevator be located in the elevator shaft of a building on Earth and assume that for the moment it is at the top of the shaft and is at rest.

The observers in these two elevators will both find that their ping pong game goes as usual. The observers in the elevator being accelerated upward in space will find that the floor of the elevator presses upward on the soles of their feet with a force equal to their Earthly weight. These observers will be unable to establish by any experiment that they are not at rest on the surface of the Earth (neglecting any small effects due to Earth's rotation).

Now let's put the fox amongst the chickens. Imagine that the elevator in space ceases to be accelerated and the elevator at the top of the shaft back on Earth has its cable cut. For the regretfully short time until the elevator impacts the bottom of the shaft, the occupants of the Earth-bound elevator will find that their gravitational field seems to have disappeared just as though they were in outer space. This is, of course, due to the fact that the elevator and its occupants (and the ping pong equipment) are all falling at the same rate. The observers in the space-bound elevator that is no longer accelerating will have the same sickening sense of falling as their unfortunate colleagues on Earth, however, they will have much longer to adjust to it. Neither sets of observers will be able to establish by any conceivable experiment that their situations differ.

In terms of the Eötvös experiment, the principle of equivalence amounts to the assumption that the ratio of inertial mass to gravitational mass is the same for all substances and that this ratio is equal to one. Thus far, no experimental facts have come to light to suggest that this is not so. Well, you may say, this is interesting, but not extraordinary. Don't go away; extraordinary coming right up.

Let us apply Einstein's principle of equivalence to the situation shown in Fig. 5-3. Here we have two observers, call them I and II. In Fig 5-3(a), the two observers, separated by a distance ℓ, are displaced vertically in a uniform gravitational field whose acceleration is g. You might imagine one of them to be at the top of a mine shaft, and the other at a depth ℓ. In Fig. 5-3(b), the two observers are located in a noninertial reference frame whose upward acceleration is equal to g. It's like the elevator thought experiment all over again. According to the principle of equivalence, these two situations should be completely indistinguishable and any experiments performed in either of them should give the same result.

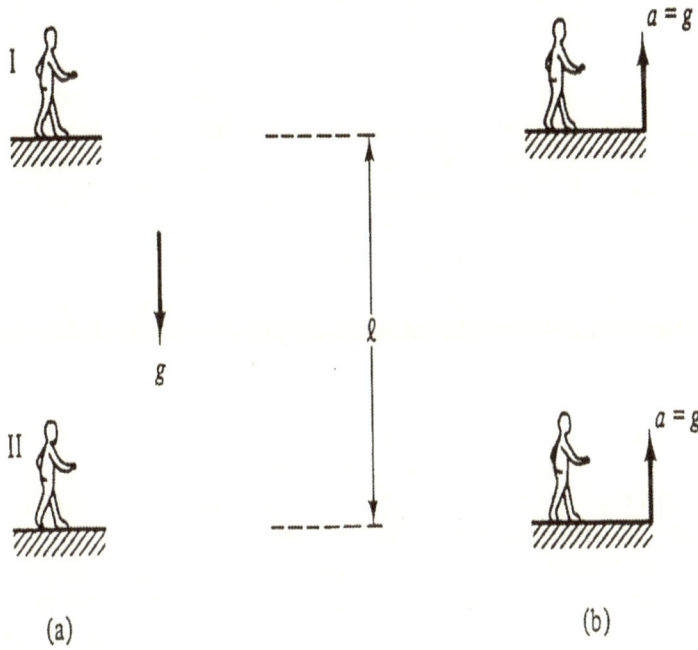

Figure 5-3 Equivalent noninertial frames of reference.

Suppose now that observer I in Fig. 5-3(b) flashes a pulse of light of frequency v in the direction of observer II. It can be shown that owing to the time of approximately ℓ/c that it takes the light to travel from observer I to observer II, observer II will have acquired a velocity $g\ell/c$ in excess of the velocity that observer I had at the time the light was emitted. This apparent relative velocity between the two observers comes about owing to the fact that they are both accelerating and that the light takes a finite time to travel from I to II. This results in a Doppler shift in the observed frequency of the light seen by II. It will be somewhat bluer than the light that left I. The Doppler effect (in case you don't know and have not lately been caught speeding by Police radar, which works on the Doppler principle) is the apparent crowding together of wavefronts owing to the motion of the receiver toward the source. This effect gives the received wave a higher frequency than it had when it was emitted. In this case, the fractional change in the frequency $(v' - v)/v \approx g\ell/c^2$ where v is the emitted frequency and v' is the received frequency. There is nothing at all spooky about this result. It is based on simple Newtonian mechanics and the Doppler effect. If we now invoke the principle of equivalence, we see that this result derived for the situation depicted in Fig. 5-3(b) must be equally applicable to situation (a), where the accelerated frame of reference is replaced by a gravitational field whose acceleration g is numerically equal to the acceleration of the reference frame in (b). This implies a highly non-Newtonian result, namely the interaction of light, a massless wave phenomenon, with the gravitational field. On Earth, for a fall of 1000 feet (305 meters), the fractional frequency shift is only 3.3 x 10^{-14} (about 3 parts in a hundred million million for those of you unused to powers of ten notation). As small as that is, a technique called the Mössbauer effect allows it to be measured. In 1959, R. V. Pound and G. A. Rebka, Jr. measured the gravitational shift of a beam of light allowed to "fall" down a 144-foot mine shaft. The ratio of the experimentally measured result to the theoretically predicted result for Δv/v was 1.05±0.10. By 1965, the experimental accuracy had improved to 0.999±0.0076. The effect cannot be doubted. Gravity does indeed act upon light.

The Principle of Equivalence can also be interpreted to imply that clocks run more slowly in a gravitational field. This has been measured and confirmed. Here on Earth, this time dilation effect is infinitesimal, but in the gravitational field of a neutron star or a black hole, the effect is significant.

There is still more insight to be had from the principle of equivalence. We saw that, as a consequence of this principle, light can be shown to interact with the gravitational field. As we shall see, this interaction manifests itself not only as a spectral shift but also by the deflection of a beam of light moving perpendicular to a gravitational field. To show how this comes about, let's postulate another thought experiment featuring elevators.

As before, we suppose that one of the elevators is located in deep space, far from any source of gravitation, and is being drawn upward at an acceleration equal to the acceleration of gravity at the surface of the Earth. The other elevator is at rest on Earth (don't worry; we're not going to kill its inhabitants this time). According to the principle, these two situations are equivalent.

Now let the observers in the space elevator turn on a light source on one wall and cause a beam of light to be emitted horizontally. A moment's thought will show that this light beam must hit the opposite wall at a point somewhat below the point from which it was emitted. This is due to the fact that during the finite time the light took to cross the elevator, the floor of the elevator, being under acceleration, was rising to meet it. Hence the observer in the space elevator will see the light beam appear to curve toward the floor and since he knows his situation, this elicits no surprise. According to the principle of equivalence, the observer in the terrestrial elevator must also find that his light beam curves downward, however here, the gravitational field must appear as the cause. How much of a dip are we talking about? If ℓ is the width of the elevator and g is its upward acceleration, then the drop distance of the beam can be worked out by a simple Newtonian calculation to be ½gℓ^2/c^2, which amounts, on the surface of the Earth to about five billionths of an inch per mile of beam travel, quite ignorable by any standard. As with the gravitational time dilation, we see that these effects are only significant on the cosmic scale (very large distances) and/or for very intense gravitational fields.

This question of the path taken by a beam of light is a fundamental one in that it is intimately associated with our perception of space. When we look into a convex or concave mirror at the carnival fun-house, we see an image that appears distorted in comparison with the normal Euclidean spatial relationships with which we are attuned. Thus, we can say that space, as we perceive it, appears to be non-Euclidean in the presence of a gravitational field.

There is an additional source of spatial curvature that has nothing to do with the principle of equivalence, but can be predicted on the basis of special relativity. In Fig. 5-4, we show a disc that is rotating at a constant angular frequency ω. The circumference of this disc, and also its radius, have a large number of very small rulers placed on them. We can measure the circumference and the radius by counting these rulers and can approximate an inertial frame of reference for the application of special relativity by considering very short times. The rulers placed along the circumference will be subject to Fitzgerald contraction since they are aligned in the direction of motion. The rulers laid along the rotating radius will be unaffected, however, since they are aligned perpendicular to the motion. The result is that an observer rotating with the disc will continue to see the disc as flat and Euclidean with the traditional ratio of circumference to radius of 2π. For the stationary observer, the circumference appears contracted while the radius does not. Thus the ratio of circumference to radius is now observed to be $2\pi/\gamma$, where γ, the dilation factor that governs clocks and rods, is one for $\omega = 0$ and tends to infinity as the speed of the circumference approaches the speed of light. The consequence is that a rapidly rotating disc appears to turn into a bowl.

We need to carry this discussion of the relationship between spatial geometry and the path of light beams somewhat further. The nature of three-dimensional Euclidean space is equivalent to the statement that the shortest distance between two points is a straight line. If we turn our attention to the Minkowski space of special relativity (a four-dimensional Euclidean space), we find that straight-line paths in Minkowski space are also extreme. Instead of being the shortest distance between two points, however, we find that a straight line in Minkowski space is the longest path between two events (we refer to points in Minkowski space as events).

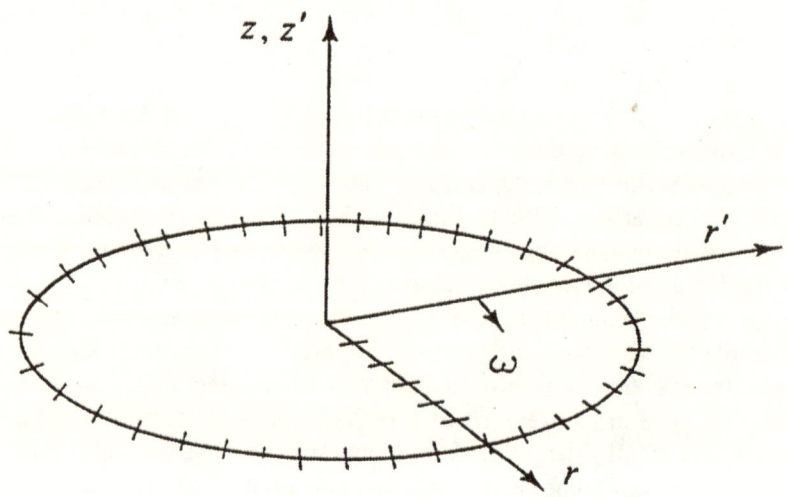

Figure 5-4 A circular disc rotating with respect to the primed frame and at rest with respect to the unprimed frame. The circumference has arbitrarily short measuring rods attached to it as does the radius r that rotates with the disc.

Extremal paths in a space are referred to as *geodesics* and there is a very special relationship between these geodesics and the path followed by a beam of light. This relationship is specified by Fermat's principle, which states that *a light ray, in traveling from one point (or event) to another, will follow a path such that the time required is a maximum or a minimum or remains unchanged as compared with propagation by any neighboring path.* Thus, not all geodesics are light paths, but all light paths are geodesics. Einstein himself made the point that space is what we measure with rods and time is what we measure with clocks, nothing more. Thus we are encouraged to regard the bending of light beams and dilation of time as reflecting the curvature of space-time itself owing to its contents of mass, energy, and momentum. This is the sort of reasoning that led Albert Einstein to formulate the field equation of general relativity.

Newton's law of gravitation implies instantaneous action-at-a-distance, and that sort of non-locality would not do for Einstein. The new theory would have to reduce to special relativity and to a local form of Newton's law of gravitation in the limit of weak gravitational fields. Also, Einstein desired that it include Mach's principle so as to account for inertial forces in geometrical terms. In order to meet these requirements and some additional mathematical ones, he found that the curvature of four-dimensional space had to be described by a *tensor*, and not just any tensor. A short (painless) lesson in mathematics follows.

Some phenomena like temperature or mass can be described by just one number, and that number is called a tensor of the zeroth rank or a *scalar*. Other quantities require three descriptive numbers, such as the x, y, and z components of a force in ordinary three-dimensional space. Such quantities are referred to as tensors of the first rank or *vectors*. There are other situations such as stress and strain that require nine numbers to specify them; these nine-number sets are referred to as tensors of the second rank. Normally, that's about as gnarly as it gets. The curvature tensor needed to describe the geometry in general relativity is a four-dimensional tensor of the fourth rank, which calls for 256 measure numbers (the number of measure numbers is the dimensionality raised to the power of the rank). By all reasonable accounts, this makes the curvature tensor the most magnificent mathematical object in all of physics. When all of the symmetries are included, this tensor boils down to 20 independent components.

In 1915, Einstein published his field equation. It contains literally worlds of information, but the equation is non-linear and the information is not easy to extract. The left-hand side of the equation describes the geometry of space-time. The right-hand side describes the stress-energy-momentum content of space-time. The two terms are related by a constant such that the equation reduces to the familiar laws in the Newtonian limit.

For other than cosmological distances and very large gravitational fields, the effects of general relativity are extremely small and it is not easy to carry out experimental tests. As luck would have it, Karl Schwarzschild managed, in 1916, to find an exact solution for the space-time surrounding a spherically symmetric mass.

As you may recall, Kepler and then Newton had shown that the orbits of planets about the Sun are ellipses. Since the mid-nineteenth century, astronomers had been aware that the orbit of Mercury was not quite perfect, even after the perturbing effects of the outer planets had been taken into account. The elliptical orbit appeared to rotate slowly (*precess* is the word astronomers use). As a result, the orbit of Mercury is a rosette figure instead of a perfect ellipse.

Calculations carried out using Schwarzschild's solution of Einstein's field equation included a term not present in the Newtonian orbit equations, which had the effect of mandating such a precession. The effect would be very small, but it would be larger for Mercury, the innermost planet than for any other planet. General relativity predicts a precession of 43.0 seconds of arc per Earth century. The measured value for Mercury is 42.6 ± 0.9 seconds per century, an entirely satisfactory agreement.

Another test of general relativity is provided by the prediction that light, in passing close to a massive body, will be deflected as shown in Fig. 5-5. In order to calculate the deflection, you have to do a little more than we did when we came up with a Newtonian estimation for the deflection of light in an accelerating elevator. The general relativistic equation for the geodesic for light must be solved. In this way, one calculates that the angular deflection of starlight that just grazes the limb of the Sun is 1.745 seconds of arc

In order to verify this figure, it was necessary to make observations during a solar eclipse. World War I delayed the experiment until 1919. On May 29 of that year, an eclipse was due and fortunately, the Sun was then located in line with a very rich field of bright stars, part of the group known as the Hyades. Two expeditions were sent out by the Royal Astronomical Society, one headed by Crommelin to Sobral in northern Brazil, and the other headed by Sir Arthur Eddington (1882-1944) to the Isle of Principe in the Gulf of Guinea, West Africa. Measurements were obtained from both locations. The Brazilian expedition measured a deflection angle of 1.98±0.12 seconds and the Principe expedition measured 1.61±0.30 seconds. These measurements have been repeated at each eclipse since then and the agreement with the predictions of general relativity is now better than one part in 500.

For many years, there were only these two tests, and except for a few people who were busy looking into the cosmological implications of general relativity, the subject was little pursued. Several events in the 1960s, notably the space program, served to rekindle interest until today, general relativity and cosmology are hot topics.

Our ability to make tests of general relativity have improved considerably over what they were 80 years ago. Irwin Shapiro and his team at MIT have bounced radar beams off inner planets such as Venus and Mercury just as they were about to be eclipsed by the Sun. By measuring the transit time for the beam, they were able to calculate very accurately the angle through which the radar beam was bent by the Sun's gravity. These results were in very close agreement with the predictions of Einstein's theory.

The most dramatic example of the gravitational deflection of a light beam was discovered in 1979. If light from a distant object happens to pass very near to a really massive object on its way to Earth, then the light can be substantially deflected, much as light is bent (refracted) in a glass lens. Einstein predicted such a gravitational lens effect in 1937. He showed that if such a gravitational lens was found that we would see a multiple (most generally, double) image of the distant object emitting the light. In 1979, Dennis Walsh, Robert Carswell, and Ray Weymann noticed a double image of a distant quasar in their telescope. This was shown to be a gravitational lens effect caused by a galaxy lying on the line of sight between us and the distant quasar. We shall have more to say about gravitational lenses later on.

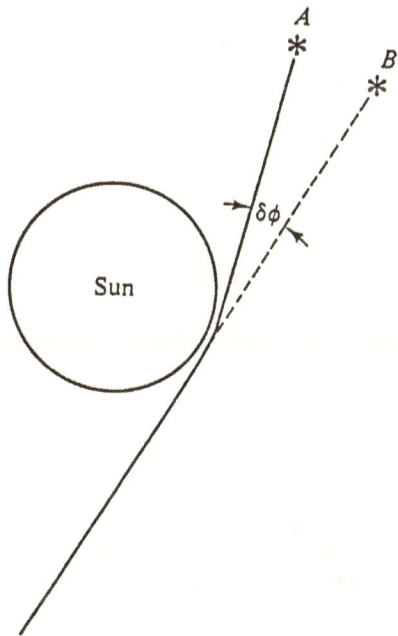

Figure 5-5 The deflection of light from a star located at A. As the light passes close to the limb of the Sun, it is bent through an angle δφ. As a result, the star appears to be located at B.

Another prediction that general relativity makes is that of gravitational waves, oscillations of the curvature of space-time that propagate at the speed of light. Joseph Weber at the University of Maryland built a gravitational wave detector and announced in the late 1960s that he had succeeded in detecting them. Others were unable to reproduce Weber's data, however. More recently, careful measurement of the periods of rapidly rotating neutron stars (pulsars) over a number of years have shown that the rate of their rotation is slowly decreasing. Pulsars, particularly binary pulsars, may be expected to be powerful emitters of gravitational waves. These waves carry off energy, which must come at the expense of the rotational energy of the pulsar. This is indirect evidence of gravitational waves, to be sure, but it is highly suggestive.

There is, of course, also the evidence of the gravitational spectral shift, but this really applies only to the principle of equivalence and not to general relativity as a whole.

The thing that has really caused the subject of general relativity to take off is the heightened interest in black holes and cosmology. These are both such current topics that to try and say very much about them as someone who is not active in the field would be to risk giving you outdated and possibly misleading information. For this reason, I will touch on both of these subjects only lightly.

The Schwarzschild solution to the general relativistic field equation predicts some sort of mathematical catastrophe (what the mathematicians call a *singularity*) as the distance of approach to a gravitational source comes close to the value $2GM/c^2$, where G is the Newtonian gravitational constant. This value (the event horizon) corresponds to an approach to the point where the gravitational field of the object is so strong that even light cannot escape it. For an object of one solar mass, this distance is just under two miles (about three kilometers).

It appears quite likely that black holes do indeed occur in nature. Likely candidates have been detected. If the burned out core of a star that has depleted its nuclear energy resources is much more than three solar masses, then there is no mechanism that can prevent its complete gravitational collapse.

Following collapse to a static (Schwarzschild) black hole, all information about the collapsed object is irretrievably lost except for its mass, which still manifests itself in the gravitational field of the hole. In the years between 1916 and 1918, H. Reissner in Germany and G. Nordstrøm in Denmark worked out the solution for an electrically charged black hole. The Reissner-Nordstrøm black hole has two properties, mass and charge. Such holes are of little practical interest since an initial charge would be neutralized rather quickly by opposite charges sucked in by the combined gravitational and electrical fields.

Of much greater interest was the solution appropriate to a spinning black hole derived in 1963 by Roy P. Kerr, a New Zealand mathematician working at the University of Texas. Since all celestial bodies are observed to rotate, and the angular momentum corresponding to this rotation will be conserved in the collapse, this solution is of the highest interest. We would expect all naturally occurring black holes to be Kerr black holes.

The solution for a static black hole predicts a single event horizon from which the escape of light is no longer possible, and a point singularity of mass at the center. The Kerr black hole is somewhat more complicated. There will be two event horizons, one associated with the mass and the other with the spin. The singularity at the center will appear in the form of a ring rather than a point.

I am tempted to go on and discuss the predictions of general relativity regarding the existence of other universes, some with repulsive rather than attractive gravity. You will be better served, however, by reading what the experts in the field have to say.

I would like to say a few words about the cosmological implications of general relativity, and then I am done. First, a little background on the problem of discussing the large-scale structure of the universe. In 1823, Wilhelm Olbers posed a question that has come down to us as *Olbers' Paradox*, although we know that it was discussed as long ago as Kepler's time. If we assume a very large static Euclidean universe, the standard pre-relativity model, then Olbers' question: "Why is the night sky dark?", is thought-provoking indeed. It's like being in the middle of a large forest where you see a tree in every direction. Why is there not a star in every tiny increment of direction in which we look, and thus why is the night sky not uniformly bright? In Olbers' time, people reasoned that dust and gases might be present to absorb the light from distant stars, but that won't do. Given enough time the gas and dust would reach the same temperature as the stars and would begin to glow as well.

There are a couple of ways out of this dilemma, both of them due to Edwin Hubble. In the 1920s, Hubble discovered that the optical spectra of distant galaxies exhibited a red shift and that this shift seemed to be proportional to the distance of the observed galaxy from the Earth. Hubble assumed that he was seeing a Doppler shift implying that distant objects were receding from us at high speed. With this assumption, which is almost universally accepted, we find that the magnitude of the red shift is proportional to the distance, and that the more distant objects are receding at higher velocities. The currently accepted value of the Hubble constant, which relates these quantities, suggests that the universe is expanding from some initially compact condition and that this expansion began about fifteen billion (thousand million) years ago.

Now back to Herr Olbers' question. If the universe is expanding, then the light from distant stars is shifted toward the red and out of the visible spectrum. That would explain the darkness of the night sky. The other explanation is that as we look out in distance, we are also looking back in time, owing to the finite speed of light. E. R. Harrison at the University of Massachusetts has calculated that the finite age of the universe gives an even more important contribution to the solution of Olbers' paradox. Most of the radiation that would contribute to a uniformly bright night sky would have to come from stars that were very distant and hence very old - some 10^{24} years. On the basis of Hubble's law, we know that the age of the universe is of order 10^{10} years, so Hubble's law and the value of his constant effectively dispose of Olbers' paradox.

When Einstein first started looking at cosmological solutions of his general relativistic field equation, he was looking for closed static solutions (this was before Hubble discovered the expansion). Einstein was forced to include an extra term (the *cosmological constant* Λ) in his field equation in order to balance the one-way pull of gravity. Eddington and Lemaitre both showed that Einstein's solution is unstable and will expand if perturbed. Einstein was very embarrassed by the error and declared that the inclusion of the extra term was the worst mistake of his career. Eddington (who had a very good intuition) was not so sure that it should be tossed out. The question of whether Λ is zero or nonzero will have to be settled by observation in the last analysis.

If $\Lambda = 0$, then three solutions are possible for the radius R of the universe as a function of time, corresponding to an open (Lobatchevskian) universe with curvature index k = -1 which expands forever, a closed (Riemannian) universe with k = 1 which eventually collapses back upon itself, and a flat (Euclidean) universe with k = 0 in which the expansion gradually slows to zero as R approaches infinity. These solutions are shown in Fig. 5-6. It is worth noting that only for closed solutions is Mach's principle found to apply. All of these cosmological models predict expansion from an initially compact state and are known collectively as *Big Bang* models.

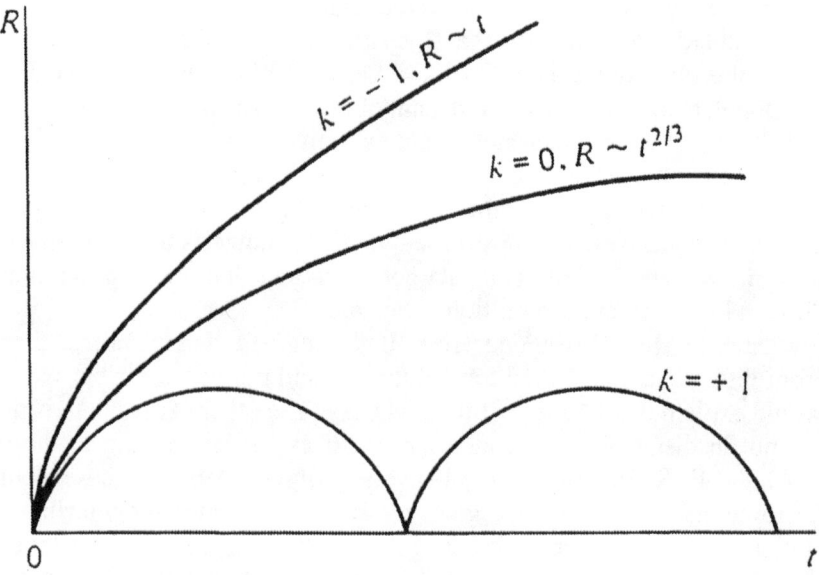

Figure 5-6 General relativistic models of the universe with $\Lambda = 0$ and finite content of matter.

The 'big bang' models used to have a serious competitor, the steady-state theory of Bondi, Gold, and Hoyle. This model extended the Cosmological Principle, which states that the Universe is homogeneous on the large scale, to a Perfect Cosmological Principle, which requires that the Universe be unchanging in time, always looking just as it does now. This was accomplished by requiring that continuous creation take place (about one hydrogen atom per cubic kilometer per year will do the trick), this creation of new matter providing the pressure that drives the expansion. This model was very appealing from a philosophical point of view.

The steady-state model was dealt a death blow in 1965 by the discovery of a general background radiation corresponding to the expansion of the Universe from a much more dense earlier state. This could, of course, only be consistent with the Big Bang model. This radiation was discovered by Arnold Penzias and Robert Wilson. It had been predicted in 1948 by George Gamow. There is general agreement as to this interpretation of Penzias' and Wilson's findings.

Twenty years ago, a direct census of stars, dust, and gas led cosmologists to believe that there was not enough matter in the universe to close it and that the expansion would continue forever. Visual census is only one way of determining the matter content of the universe, however. By measuring the relative amounts of hydrogen and helium and by studying the dynamics of galaxies and galactic clusters, astronomers have been forced to the conclusion that more than 90 percent of the matter in the Universe remains unseen. As I write this, the nature of this *dark matter* is a (dark) matter of great interest. It therefore seems quite likely that there is sufficient mass, seen and unseen, to close the Universe. The question of whether or not the expansion will continue forever or whether the Universe will begin to collapse at some point depends upon the value of Λ, the cosmological constant. The possible expansion tracks for a universe with nonzero Λ are shown in Fig. 5-7. At the time that I write this, measurements seem to indicate that the expansion is accelerating, which would argue for a finite cosmological constant. Time will tell. Our conclusion as to the Universe being closed is likely to hold up. As we remarked above, only a closed universe can display inertial effects, which are certainly a matter of common experience.

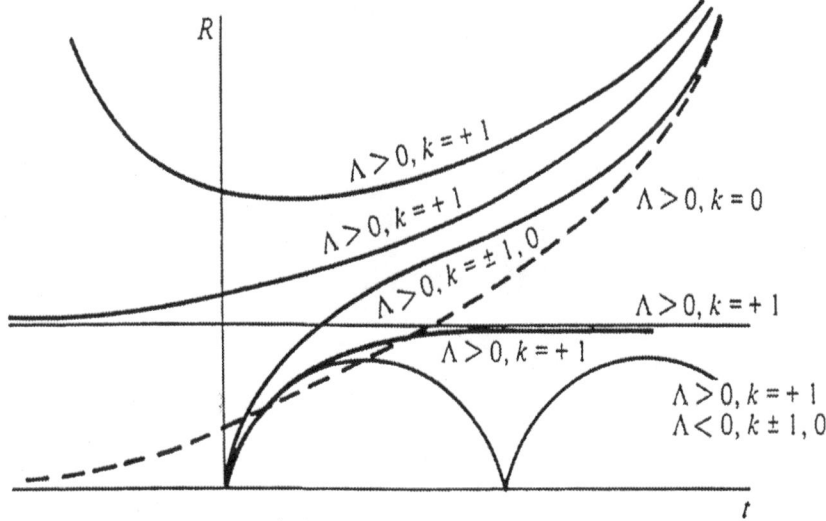

Figure 5-7 General relativistic models with $\Lambda \neq 0$ and various values of the curvature index k and average matter density.

In the last two chapters, we have described many wonders - contraction of rods, dilation of time, travel into the future, warped space-time - and yet we are still dealing with the world of appearances described by classical physics. We have simply looked at very fast, very large, and very dense phenomena that are far from our common experience. We have not yet looked upon the reality that underlies the world of appearances.

In the next chapter, we will begin to do so.

CHAPTER SIX

in which the grim reaper knocks on the
door of classical physics.

In the last years of the nineteenth century, some new phenomena began to emerge from the laboratory.

Wilhelm Konrad Röntgen discovered X-rays in 1895 while working with a cathode-ray tube operated at high voltage. He noticed that barium salts near the tube were observed to fluoresce when the tube was turned on. This fluorescence was undiminished when the tube was covered with black cardboard. Röntgen managed to establish that the radiation was originating from the point where the beam hit the glass wall of the tube, as shown in Fig. 6-1. The Röntgen rays, or X-rays as they came to be called, were found to expose photographic plates, thus providing a convenient detector. Röntgen was unable to observe anything such as refraction or interference effects that would tell him whether he was dealing with waves or particles, hence the name X-rays to denote his ignorance of their nature. Unknown nature or not, within three months of Röntgen's discovery, X-rays were being used in a Viennese hospital for presurgical diagnosis.

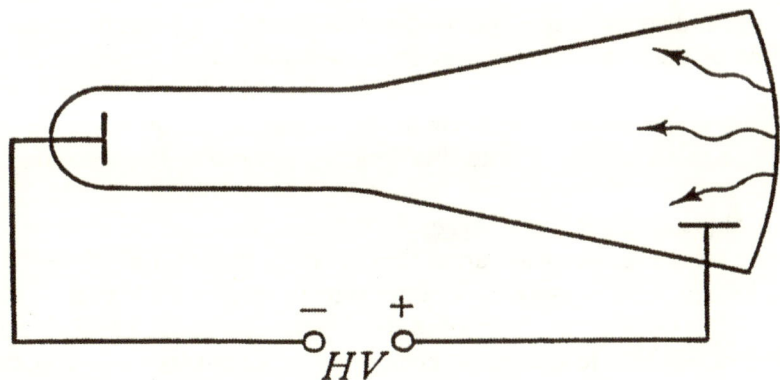

Figure 6-1 A Röntgen X-ray tube. Electrons emitted from the cathode at the left are accelerated, striking the end of the tube at the right and causing X-rays to be generated.

Nuclear physics had its beginnings in the 1890s as well. In 1896, Henri Becquerel discovered that certain salts of uranium appeared to spontaneously emit a radiation that was quite similar in its penetrating properties to X-rays. Becquerel had discovered radioactivity. In 1898, Marie Curie, working in Paris with her husband, discovered the existence of two new elements, both more radioactive than uranium. One was named polonium (for Poland, where Marie Curie was born); the other was radium. Marie Curie shared the Nobel Prize for physics with her husband Pierre Curie and Henri Becquerel in 1903, and won the chemistry prize quite on her own in 1911.

The discovery that is of primary concern to us, however, was the discovery of the electron in 1897 by J. J. Thomson at Cambridge University's Cavandish Laboratory. As we have seen, cathode ray tubes and gas discharges were all the rage in the 1890s. Thomson discovered that the visible beam in the tube (visible because the electrons excited the gas molecules in the rather poor vacuum inside the tube) could be deflected by the application of transverse electric and magnetic fields. This suggested that the beam was composed of charged particles, whose charge-to-mass ratio Thomson was able to measure. The word *electron* had been proposed earlier for a discrete unit of charge and this term was applied to the cathode ray particles discovered by Thomson.

The physics community was strongly divided into those such as Ernst Mach who did not believe in atoms as anything other than conceptual constructs (if that) and those who thought they had objective reality. The latter group were busy speculating on atomic structure. Thomson favored what came to be called the *plum pudding* model wherein electrons were embedded in a spherical background of positive charge, the atom being neutral overall. If you are American, and are not quite sure what a plum pudding is, think fruitcake.

None of these new discoveries threatened the towering edifice of classical physics. The research that did introduce that first bit of grit into the gears of the clockwork universe had as its object the most mundane of questions. Why does a heated fire poker glow red instead of blue, as it should? The reasoning goes as follows.

Hot objects radiate electromagnetic radiation because the electrons in their atoms are set to oscillating as part of the overall thermal motion of the molecules. It was known from Maxwell's electromagnetic theory that when a charge is accelerated, it radiates. Therefore one should expect a charge oscillating about its equilibrium position to radiate an energy per unit time (power) proportional to the square of its oscillation frequency. A Thomson "plum pudding" atom with radiating electrons is shown in Fig. 6-2. Blue light corresponds to a higher oscillation frequency than red, and so most of the energy of the radiation should be expected to be concentrated at the blue end of the spectrum. For this reason, one would expect all heated objects to glow blue, the hotter, the brighter, but blue at all temperatures. Nature does not accommodate this straightforward reasoning at all. Thermal radiation enters the visible portion of the spectrum from the red (low frequency) end. As you continue to heat the object to higher temperatures, it gets brighter as you would expect and the peak of the spectral distribution moves toward the blue (higher frequencies). In the late 1890s, quite a few people were trying to sort out this annoying little problem.

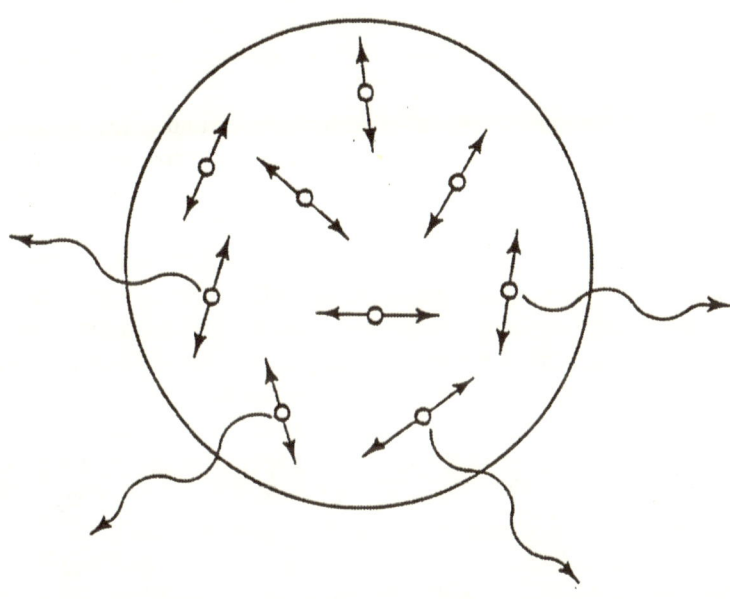

Figure 6-2 The Thomson model of a multi-electron atom showing electron oscillation and consequent radiation.

Lord Rayleigh had followed just this sort of reasoning in deriving an equation for the spectral distribution of thermal radiation. Sir James Jeans contributed to this work and the distribution law bears both their names, the Rayleigh-Jeans law. It agrees with the experimentally measured distribution on the low-frequency end, but fails spectacularly at high frequencies, predicting infinite radiative power at high frequencies in what was called at the time the *ultraviolet catastrophe* (sounds like a good name for a rock group). Max Wien, using a thermodynamic approach, derived a different spectral distribution law, published in 1893. The Wien law showed good agreement with experiment at high frequencies, but failed at low frequencies. So when Max Planck got involved with the problem, there existed two theories of what was known as *blackbody radiation*, each only half right.

Planck took the Rayleigh-Jeans approach, but where they had taken an expression for the average energy per oscillation mode based on the thermodynamic principle of equipartition of energy, Planck tried something else. He knew he had to find a mechanism that would suppress the energy radiated at high frequencies so as to avoid the ultraviolet catastrophe. In order to do this, he assumed that a one-dimensional oscillator could only have energies proportional to an integral multiple of the frequency. In other words, $E_n = nh\nu$, where n is an integer (n = 0,1,2,3,..), ν is the frequency of the oscillator, and h is the constant of proportionality. There was no physics behind this assumption. It was merely a case of "suck it and see." If it worked, Planck had in mind taking the limit as the constant of proportionality, h, approached zero and thus recovering the continuous spectrum that everyone knew must characterize classical oscillators. To make a long story short, Planck found that when he let h go to zero, he recovered the ultraviolet catastrophe, but if he gave h a specific non-zero value, then he could fit the experimental curve. In the process, Planck not only found an accurate value for h (which has come to be called Planck's constant), he also determined the charge on the electron more accurately than anyone had up to that time. Planck published this work with some personal misgivings in December of 1900. It was too radical to attract much immediate attention, except from Einstein, whose nose for significant information was very acute.

You may recall that back in the chapter on special relativity I mentioned that Einstein published two other papers in 1905 in addition to the paper on special relativity. The first one of those showed that atoms were real.

The idea of atoms had been around for over two thousand years. The Greek philosophers and Indian sages had both entertained the notion. The successes of the kinetic theory of gases (modeled as a large number of energetic molecules), striking though they were, failed to provide direct confirmation of the existence of atoms and molecules. The necessary confirmation was provided by Einstein in the form of a successful theoretical explanation of an effect first observed in 1828 by an English botanist, Robert Brown.

Brown had noticed that pollen grains suspended in water were subject to an irregular zigzaging motion. Einstein assumed that the motion, known as *Brownian motion*, was caused by multiple collisions between the pollen grains and the water molecules. On that basis, he derived an expression for the diffusion rate of the grains as a function of the temperature of the water, the size of the grains, and the coefficient of viscosity. Jean Perrin verified the correctness of Einstein's formula in 1908. Henceforth, the existence of molecules could no longer be doubted.

The third of Einstein's 1905 papers was the one that won him the Nobel Prize. In order to discuss that one, we first need to get a little bit of historical background.

It was mentioned in Chapter Three that Heinrich Hertz first generated and detected nonoptical electromagnetic waves (radio waves) in 1887. It is one of the ironies of physics that this experiment, which served to corroborate Maxwell's theory of electromagnetic waves also contained data that were consistent only with a corpuscular (particle) interpretation of electromagnetic radiation.

Hertz' apparatus consisted of an oscillating inductive-capacitive circuit with a spark gap. Electromagnetic radiation from this circuit was detected by the observation of sparks in a similar tuned circuit across the room from the first. In order to observe the weak sparks in the receiver gap more easily, Hertz enclosed the primary gap in an opaque box and found that the receiver gap had to be shortened considerably in order to obtain sparks. He further found that even the imposition of a plate of glass between the two gaps produced this effect. If the receiver gap was exposed to the unfiltered radiation from a third spark gap oscillating at any frequency, the reception was enhanced to the same extent as when the receiver gap was allowed to 'see' the transmitter gap. Hertz also found that the strength of the sparks observed in the receiver gap depended rather critically on having clean electrodes.

It was known at the time that sparks have a strong ultraviolet component in their spectrum. It seemed as though exposure to this (or any other) source of ultraviolet radiation greatly enhanced the ability of the gap to conduct electricity. Hertz did not spend any time on this puzzle; his primary concern was with the radio-frequency waves.

What was the carrier for the electricity that flowed through the gap in Hertz' experiment? It was thought at first that the air molecules might have been the carrying agent. It was soon found, however, that the effect persisted even in the best vacuums that could be made at that time. Further, P. Lenard showed in 1900 that the cathode atoms were not the carriers either. It therefore seemed as though the metal was losing some sort of negatively charged particle, the nature of which was unknown.

The particle in question was, of course, the electron found by J. J. Thomson just a few years before and for which he had measured the charge-to-mass ratio. We have also seen that Max Planck obtained a good (within two percent) value for the electronic charge e as a by-product of his work on the blackbody radiation law. Lenard had shown in 1900 that the charge-to-mass ratio for photoelectrons (the charge carriers in the gap) was the same as that found by Thomson for the cathode-ray electrons. The relevant data were mounting.

In 1912, Richardson and Compton found that the maximum kinetic energies of photoelectrons emitted from aluminum decreased as the wavelength of the illuminating light was increased. Elster and Geitel showed, in a series of extremely accurate experiments carried out in 1913-1914, that the magnitude of the photoelectric current was directly proportional to the intensity of the light used to excite the effect.

Measurements of the photoelectric effect are made in apparatus of the sort depicted in Fig. 6-3. A glass tube T is evacuated to a good vacuum. A source of monochromatic ultraviolet light is shone through a quartz window W onto a metal cathode C. A battery B is connected between the cathode and an anode A in such a way that the voltage between C and A can be varied continuously between $+V_B$ and $-V_B$. The switch S provides for the voltage reversal; with the switch in position 1, the cathode is biased positive and the electrons experience an attraction for the cathode. In position 2, the electrons are accelerated toward the anode. The variable resister R serves to vary the magnitude of the bias voltage.

In order to measure the maximum kinetic energy of the photoelectrons, the cathode is biased positive and the voltage is increased until the current tends to zero. We will denote this stopping voltage as V_0.

Richardson and Compton had found that when the cathode had positive bias, then the stopping voltage V_0 decreased with decreasing frequency v of the illuminating radiation. Elster and Geitel showed that, for a negatively biased cathode, the photoelectric current was independent of the magnitude of the bias voltage and was directly proportional to the intensity of the illuminating light. The Richardson-Compton and Elster-Geitel results are shown in Fig. 6-4, in which is sketched the photoelectric current I as a function of the voltage applied between the photoemitting cathode and the anode for (a) various intensities I of the illuminating light, and (b) various values of frequency v of the light.

If all this detail has put you to sleep, I apologize, but it's time to wake up now. I need to make some important points.

On the basis of classical wave theory, the energy content of a wave is proportional to its intensity (the square of its amplitude). If we eliminate the notion that the light serves only to trigger the release of some energy source within the metal (this notion could not be supported), then it should be expected that the current would follow the dashed curves in Fig. 6-4(a) rather than the solid ones. Clearly, classical physics was in trouble again. The ultraviolet light did not seem to be behaving as a wave

The key experiments on the photoelectric effect were carried out by R. A. Millikan in the years 1914-1916. Millikan found that if he plotted the maximum kinetic energy of the photoelectrons expressed as eV_0 as a function of the frequency of the ultraviolet light, the result was a very good fit to a straight line! This relationship is shown in Fig. 6-5.

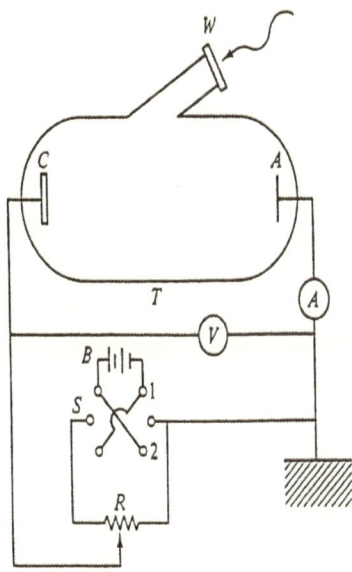

Figure 6-3 Apparatus for Observing the photoelectric effect.

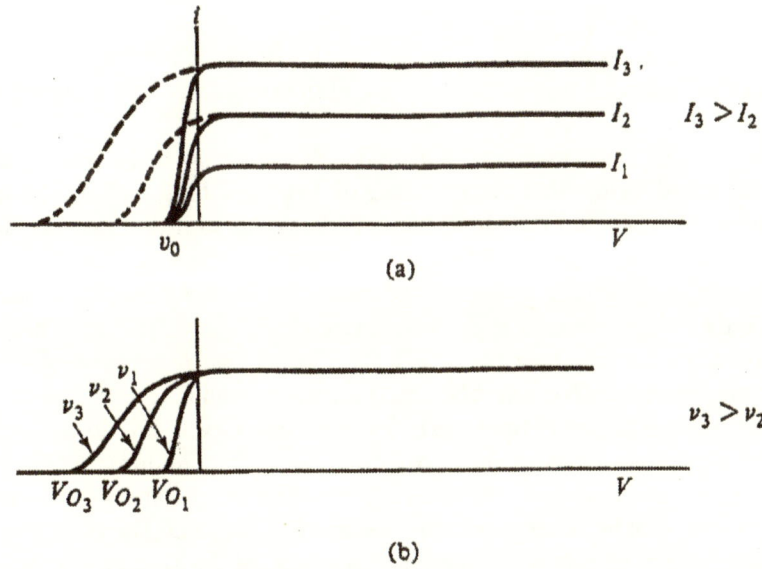

Figure 6-4 The photoelectric current (i) as a function of the voltage (V) imposed between the cathode and the anode for (a) various intensities (I) of the illuminating radiation and (b) various frequencies (*v*) of the illuminating radiation.

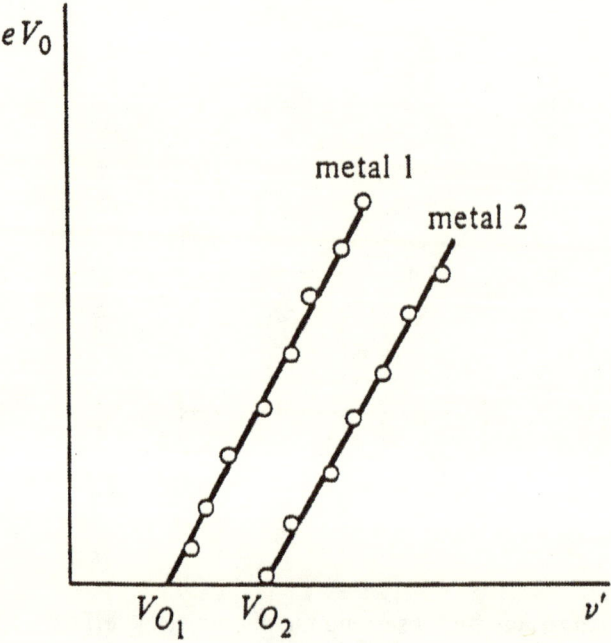

Figure 6-5 Millikan's data establishing the linear relationship between the energy of the fastest photoelectrons, eV_o, and the frequency of the light v.

This was old news to Albert Einstein. Working alone, he had predicted a theoretical relationship that fit the data of Fig. 6-5 long before Richardson-Compton or Elster-Geitel had done their work. With so little to go on, how had he done it? Einstein had taken Planck's quantization of the blackbody oscillators to heart (as almost no one had) and had made the further audacious assumption that not only is light emitted at discrete energies, $E_n = nhv$, it is also detected as little bundles of discrete energy. In effect, *Einstein had reintroduced the notion of light as a particle*! No one else on the planet would have dared do such a thing. We now call these light particles *photons*. Given this assumption, the energy of a photon, hv, goes into providing the energy required to liberate the electron from the metal (you can't simply pour them out), and what is left over goes into the kinetic energy of the electron's motion. There is the linear relationship and it couldn't be simpler. The slope of the line in Fig. 6-5 is just Planck's constant, h.

Planck had been courageous enough to publish his model of the blackbody radiation law, even though this law called for the oscillator energies to be quantized. Einstein had taken this result, known by few and believed by almost no one, and had extended the assumption to say that not only was radiant energy quantized, the radiation itself propagated in bundles having energy hv, Planck's constant times the frequency of the radiation. It was this work, published in the same volume of *Annalen der Physik* as the Brownian motion and special relativity papers that won the Nobel Prize for Einstein in 1921. Robert A. Millikan received the Nobel Prize in 1923 for his careful experimental work, both on the photoelectric effect and on the precise measurement of the electronic charge.

Newton had said that light was particles; Young and Huygens had said it was waves and Maxwell had proved it to everyone's satisfaction. Now here is Einstein saying that light can act like a particle. How can light have attributes of both particle and wave? The two are not altogether consistent. Perhaps Einstein was wrong. The photon hypothesis remained controversial for quite some time.

In 1923, Arthur H. Compton published a paper in the *Physical Review* that provided incontrovertible evidence for the existence of photons. The experimental data showed that when monochromatic (all one frequency) X-rays are scattered by passing through a graphite block, the scattered radiation contains, in addition to the primary frequency, a component of longer wavelength (lower frequency). The wavelength of this shifted component is independent of the choice of scattering material; it can also be observed using metal foils, for example. The shifted wavelength depends only on the scattering angle.

This result is totally perplexing from a classical point of view. When waves are scattered by charged particles in classical theory, the incident wave is absorbed by the particle, which is set into oscillation at the frequency of the incident wave. The oscillating charge then radiates at the same frequency and thus transfers the oscillation energy that it absorbed from the incident wave to the outgoing scattered wave. There is no mechanism whatsoever for a component with a different frequency to arise in this process.

Compton solved this mystery by assuming that the X-rays behaved like photons rather than waves. Since the relative wavelength shift was found to be independent of the frequency of the incident X-ray and the nature of the scattering material, Compton inferred that the shift was due to interaction between the X-ray photons and the orbital electrons. The fact that the energies of the incident photons were a couple of orders of magnitude larger than typical electron binding energies enabled Compton to treat the electron as a free particle and to visualize the process as very much like a game of billiards. He then proceeded to calculate the collision process according to the laws of conservation of momentum and energy. The result shows that the scattered photon can be expected to have a wavelength greater than the wavelength of the incident photon by just the measured amount.

The data of Compton's experiment, carried out using a Bragg crystal spectrometer, are shown in Fig. 6-6. These data were in excellent agreement with Compton's theory. Photons were here to stay, but, of course, so were light waves. The situation for those who need to see "how the little men were running around" was getting quite uncomfortable. Classical physics, describing as it does the world of appearances, usually lends itself to being visualized in terms of common-sense analogies. Light as both particle and wave was difficult, if not impossible to relate to experience.

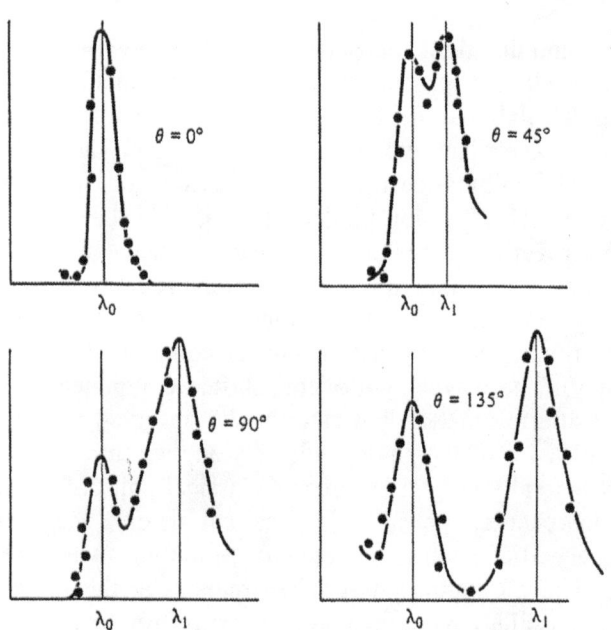

Figure 6-6 Compton's data showing the shift to longer wavelengths of the scattered X-rays. These wavelength shifts are in excellent agreement with Compton's theory. (From A.H. Compton, Phys. Rev., 22, 409 (1923).)

It has been pointed out that Albert Einstein did most of his work on the basis of instinct. The accuracy of that instinct made him quite unique in twentieth century physics. In1924, another man came along who also had an uncommonly accurate instinct and who probably deserves to be called the father of quantum physics.

Louis Victor, Prince de Broglie, was a young French nobleman who was just finishing his doctorate in 1924. He had a revolutionary idea and he had insisted on including it in his doctoral dissertation. Normally, a dissertation is the last place in the world to propose something of an *avant garde* nature. The idea is to do a solid and conservative bit of physics, get your degree, your 'ticket to live in the world', and then (and only then) do physics in whatever style takes your fancy. It helps, of course, if you have picked your parents carefully and are of independent means. De Broglie had picked his parents very carefully indeed.

De Broglie reasoned that if light had been found to manifest as particles (photons) under certain circumstances, then perhaps massive objects, which we normally observe in corpuscular form (electrons, bowling balls, planets) might also have an associated wave nature. He assumed that to an observer at rest with respect to the object, the rest energy of its particle aspect, mc^2, must be equal to the energy of its wave aspect, hv. By subjecting the standing wave to a relativistic transformation, de Broglie found that the wavelength of the associated wave is proportional to the reciprocal of the particle's momentum, the constant of proportionality being Planck's constant, specifically

$$p = h/\lambda$$

A qualitative picture of the de Broglie wave representing a particle is shown in Fig. 6-7.

De Broglie's idea came as a bolt out of the blue and not in response to some vexing problem. It did make an immediate contribution to the understanding of atoms, however. To see why this was so, I need to back up and give a little history of the development of atomic theory.

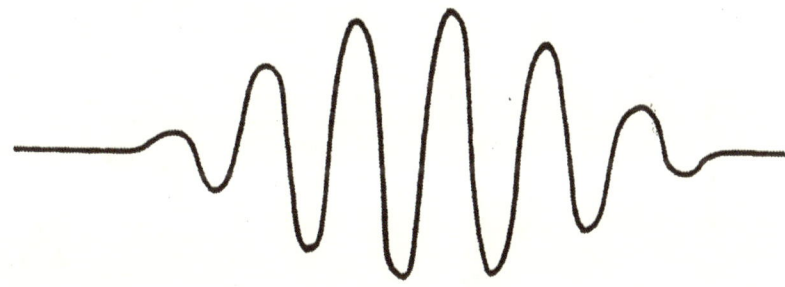

Figure 6-7 A qualitative picture of the de Broglie wave packet representing an object of finite size.

As we saw, Thomson's plum pudding model was the leading contender at the turn of the century. In 1910, two of Ernest Rutherford's students, Geiger and Marsden, carried out an experiment at the University of Manchester where they placed a thin gold foil in the path of a beam of alpha-particles (radioactive decay products later identified as energetic helium nuclei). Rutherford's purpose in this experiment was to measure the angular distribution of the alpha-particles deflected from their paths by the gold atoms. If the plum pudding model had been correct, one would have expected little or no scattering of the alphas. It would be like shooting a cannonball through a cloud. The cloud may have a mass of several tons, but there is simply not enough mass encountered by the cannonball along its trajectory to cause notable deflection.

Most of the alpha-particles were indeed scattered through very small angles. Less than one percent had scattering angles greater than three degrees. There were a few, however, for which the scattering angles approached 180 degrees. A plum pudding atom could not cause that! Rutherford realized that the positive charge of the atom must be concentrated in a small compact nucleus at the center of the atom and the electrons must circle around this nucleus much like planets circling the Sun. Rutherford published this nuclear model in 1911. Rutherford's model had an unanswered question associated with it, however. It had been known from the time of Maxwell that accelerated charges radiate. Motion in a circular orbit constitutes acceleration. Thus the nuclear model should have been incapable of equilibrium; the electrons should have radiated away their energy in about a hundred billionth of a second and spiraled into the nucleus. Clearly, this was not happening. Why?

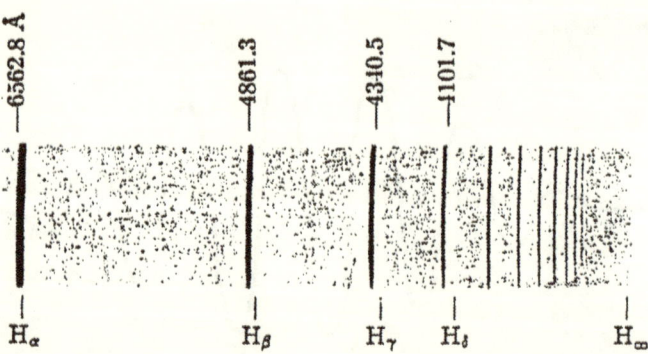

Figure 6-8 The Balmer line spectrum of atomic hydrogen. H_∞ indicates the theoretical position of the series limit. Wavelengths are given in Ångstroms (10^{-10} m = 1 Å)

The next piece of the atomic puzzle was supplied by Niels Bohr, a young Danish physicist, in 1913. Bohr set out to pick up where Rutherford had left off and model the hydrogen atom in detail. He assumed the Rutherford model with one electron in a circular orbit about the positive nucleus, both taken to be point charges, the classical Coulomb force providing the attraction. It was known that atoms only radiated discrete frequencies. The visible emission spectrum for hydrogen, the so-called Balmer series, is shown in Fig. 6-8. Bohr assumed that these discrete emission lines arise owing to the electron making a transition from an orbit characterized by an energy, say E_2, to an orbit whose energy is E_1, such that $E_2 - E_1 = \Delta E = h\nu$, where ν is the frequency of the emitted photon. Bohr's model of the hydrogen atom is shown in Fig. 6-9. By assuming that only certain quantized orbits are possible, Bohr solved the problem of accounting for the discrete line radiation of atoms, without addressing directly the question of why the atom was stable. Knowing what he wanted, Bohr went on to make whatever assumptions would produce it. In so doing, he was able to show agreement with the measured spectrum of hydrogen and with Rutherford's measurements., which had shown that the electron orbits are about one Ångstrom (0.1 nanometer) in diameter. The model was a hodgepodge of classical and quantum concepts and had no time-dependence with which to predict relative line strengths, but there were clearly some elements of truth there. The implied quantization relation for the orbits were equivalent to saying that the angular momentum of the nth orbit must be given by $L_n = pr_n = nh/(2\pi)$. If we look at this relation in terms of the de Broglie equation for the wavelength of an electron having momentum p, then we see that the Bohr quantization relation is equivalent to requiring that each orbit contain in its circumference a whole number of de Broglie wavelengths. In other words, $2\pi r_n = n\lambda$. This is depicted in Fig. 6-10. Thus de Broglie's hypothesis provides a far more fundamental basis for Bohr's quantization relation, and the success of the Bohr model provided a boost to de Broglie's otherwise unsubstantiated notion of matter waves. Still, the agreement could have been fortuitous.

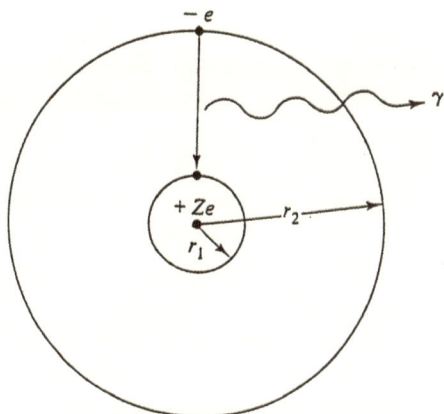

Figure 6-9 The Bohr model of a hydrogenic atom.

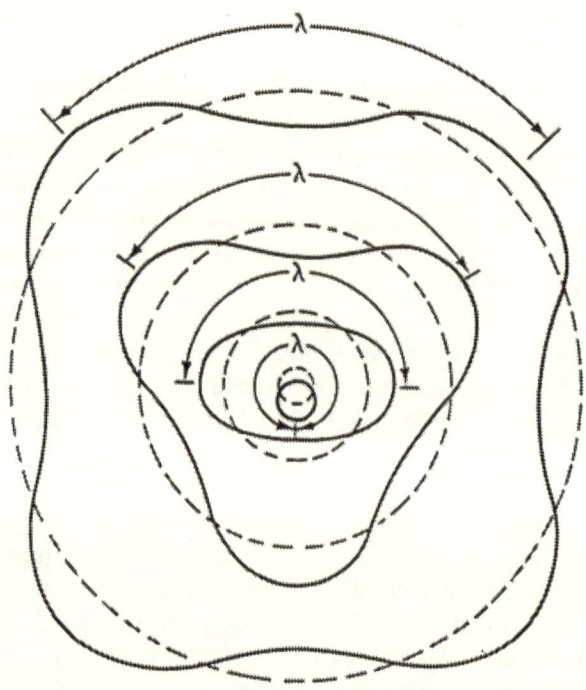

Figure 6-10 The standing-wave patterns produced by the de Broglie waves associated with electrons in the first four circular Bohr orbits of hydrogen.

Decisive confirmation of de Broglie's ideas was obtained partly by accident, as is so often the case, by C. J. Davisson and L. H. Germer at the Bell Telephone Laboratories in 1927. Davisson and Germer had been involved since 1919 in a program of study on the intensity spectrum of electron beams reflected elastically from various materials. On the particular day of their discovery, they were bombarding a nickel target with a low-energy electron beam. There was an accidental loss of vacuum that resulted in the formation of an oxide coating on the sample. Since their reflection work required a pure nickel surface, Davisson and Germer baked the target at high temperature to remove the oxide. In so doing, the nickel sample became crystallized, which is to say that its atoms were rearranged from an amorphous state into a regular array, much like assembling a crowd of soldiers into ordered ranks on a drill field. When the reflection experiments were resumed, a strong peak in the intensity of the scattered electrons appeared at a scattering angle of 50 degrees.

The experimental arrangement is shown in Fig. 6-11. The electrons were emitted by a hot wire cathode C and accelerated through a biased grid G, and onto the nickel sample. The reflected beam was detected by a collector that was biased negatively to a voltage only slightly less than the accelerating voltage on the grid. This ensured that only electrons that had undergone surface interaction would be detected, since any electrons that had penetrated the sample to an appreciable extent would have lost too much energy to overcome the negative bias on the detector. Experimental physics lives or dies by attention to details like that.

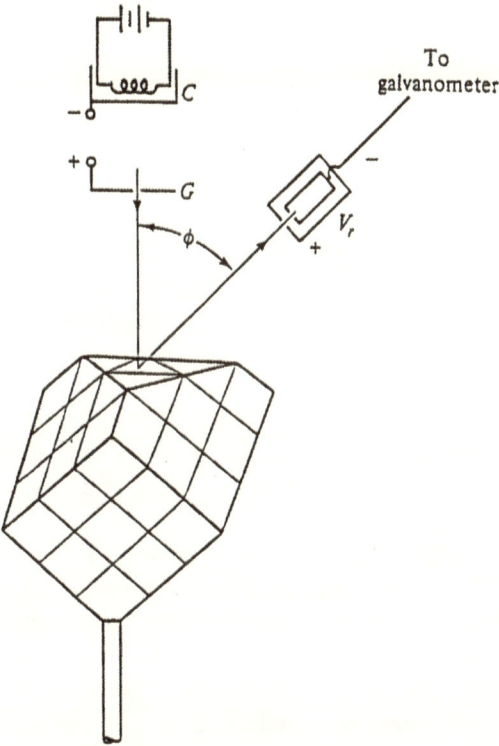

Figure 6-11 A schematic representation of the experimental layout of Davisson's and Germer's electron scattering experiment.

A polar plot of the intensity of the scattered electrons as a function of the reflection angle is shown in Fig. 6-12 for the 54 eV electrons that were used in the first experiment. (An electron-volt is the energy acquired by an electron in moving through a potential difference of one volt). A strong maximum was noted at a reflection angle of 50 degrees. The increase in intensity for smaller angles simply indicates direct backscatter.

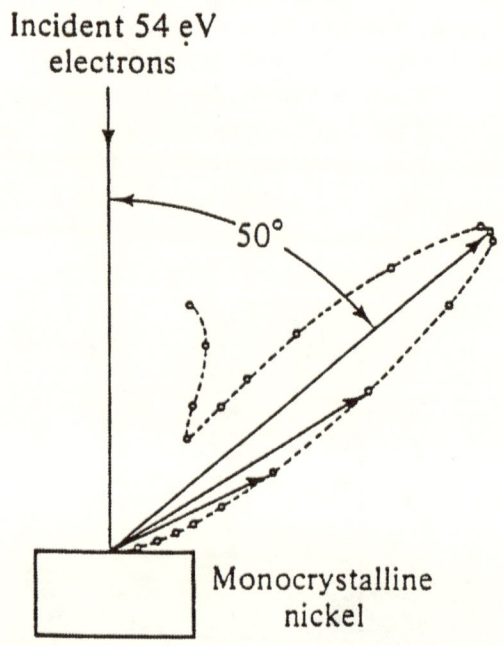

Incident 54 eV
electrons

50°

Monocrystalline
nickel

Figure 6-12 A polar plot of the relative number of electrons scattered from monocrystalline nickel for an incidnet beam energy of 54 eV

Davisson and Germer were not slow in making a correct interpretation of these results. In 1925, the year after de Broglie's thesis was published, Elsasser had pointed out that diffraction effects (a wave phenomenon resulting from multiple interference) should be measurable if an electron beam is directed onto an appropriate grating. This suggestion escaped Davisson and Germer's notice; however, by 1925 their work on elastic scattering from various materials had begun to involve them in scattering from crystals of known orientation. In the summer of 1926, Davisson learned of Elsasser's idea from Franck and others in England. Although the diffraction peak (for that was what it was) that he and Germer observed in 1927 was accidental, they were able to make an immediate correct interpretation.

Diffraction effects by electron beam transmission through polycrystalline foils were observed by G. P. Thomson (the son of J. J. Thomson, the discoverer of the electron) in 1928. He used a high-energy electron beam directed through a very thin gold foil. The diffraction ring pattern generated in this way was in excellent agreement with de Broglie's formula for the wavelength of the associated matter waves.

In the 1930s, Estermann, Frisch, and Stern succeeded in producing diffraction effects by scattering helium atoms from lithium fluoride crystals.

And so we come to the end of this chapter with confidence that light, which we ordinarily think of as a wave, also has a particle aspect, and massive matter, which we could hardly conceive of as other than corpuscular, has a wave aspect. Particles are local; waves are not. Particles are things. They occupy a definite location in space. Waves are not things. They are moving disturbances, modes of behavior. Clearly the clockwork machine of the objective deterministic world of appearances seemed to be developing a few ominous creaks and groans.

CHAPTER SEVEN

THE MADNESS GETS WORSE

in which several young men and a
couple of older ones conspire to do away
with determinism.

Werner Heisenberg was born in Würzburg, Germany in December, 1901. In
1911, the family moved to Munich, where the father, August Heisenberg, took up
an appointment as Professor of Greek Philology at the University of Munich.
Young Werner excelled not only at school, but also as a skier. He did his war
service (fortunately not in the trenches) and spent some time after the war hiking
around Germany. In 1920, he entered the University of Munich. His aim initially
was to study mathematics, but he was advised that he was already too old (at the
age of 19) for that. Heisenberg turned to theoretical physics and began to study
with Arnold Sommerfeld. He couldn't have been in a better place at a better time.
Sommerfeld had brought the Bohr atom to its final refinement, and he was
generally acknowledged to be the best teacher of theoretical physics in Europe, if
not the entire world. I very much appreciated the clarity of his textbooks, which I
had the pleasure of using as a student. One of Heisenberg's fellow students was
the abrasive, but brilliant Wolfgang Pauli, of whom more later. Heisenberg found
time to add mountain climbing to his list of skills during this time in Bavaria.

In 1923, Heisenberg received his doctorate at Munich. In his oral thesis
defense, he fell down on his general knowledge of optics, and squeaked through
with the lowest passing grade. He then moved to the University of Göttingen,
world famous as a center of mathematics, and went to work as an assistant to
Max Born. At this time (a few months before de Broglie made his matter wave
proposal) the pressure was building to formulate a *quantum mechanics*, a physics
of the atomic and sub-atomic world, to replace Bohr's crumbling model.
Nowhere was the activity more intense than in Born's group at Göttingen and
Bohr's group at Copenhagen.

During this time, Heisenberg fell strongly under the influence of the
philosophy (or non-philosophy) of *logical positivism* as set forth by Ludwig
Wittgenstein in his *Tractatus Logico-Philosophicus*, wherein he argued that:
"Whereof one cannot speak, thereof must one remain silent." Heisenberg
resolved to work exclusively with measurable quantities such as the observable
discrete spectra of atoms as opposed to nonobservables such as electron orbits. In

this mindset, he was an intellectual heir of Ernst Mach, the great positivist, who never, to his dying day, believed in atoms, never having seen one.

In late 1924, early 1925, Heisenberg spent several months in Copenhagen at the institute that the Carlsberg Brewery had built for Neils Bohr. In the spring of 1925, he was back in Göttingen, still struggling to find a toehold in the quantum mechanics problem. In early June, he came down with an acute allergic attack and left on the train for Helgoland, a barren island in the wind-swept North Sea, in order to get away from the pollen and grasses. The allergy quickly cleared up and young Werner picked up the quantum problem again. As sometimes happens after a long struggle followed by a break, the clouds rolled back and Heisenberg saw the truths that he had been seeking.

He had not solved the hydrogen atom, but he had solved the problem of a simple quantum oscillator in a self-consistent fashion. He wrote up the paper for submission to *Zeitshrift für Physik*. The paper was very difficult to understand, filled with logical broadjumps and strange ponderous infinite square arrays interspersed with familiar equations. Great importance was given to the squares of quantities, whereas the quantities themselves were held to be meaningless. Heisenberg wrote a classical expression for the electron orbit in terms of a series of sine waves (a technique pioneered by the French mathematician Fourier), but he replaced the amplitudes of Fourier terms by infinite square arrays of numbers whose rows and columns were labeled according to the initial and final energy levels characterized by the observable spectral lines. He devised rules for multiplying one infinite array by another and found that when, for example, he multiplied a position array by a momentum array, the results did not commute. In other words q times p was not equal to p times q. Never before had there been the case in physics of two physical quantities whose multiplication did not commute. Think back to the arithmetic you learned in school. Addition and multiplication both commute; $a+b = b+a$ and $a \times b = b \times a$, whereas subtraction and division do not: $a-b \neq b-a$ and $a/b \neq b/a$.

The allergic attack forgotten, Heisenberg returned to Göttingen and gave the paper to Max Born. Born saw past the rough spots and realized that it was an important piece of work. Further, drawing on his formidable background in mathematics, acquired at the feet of the great David Hilbert, Born realized that Heisenberg's square arrays were *matrices*, something with which mathematicians were familiar, and that in working out the rules with which to multiply them, he had merely reinvented matrix multiplication. It was known that matrix multiplication does not necessarily commute, but such objects had never found application in physics up to that time.

Born submitted the paper to the journal with his recommendation that it be published (without which the referees might well have rejected it). He and his student Pascual Jordan (about the same age as Heisenberg) then got to work to tidy up Heisenberg's idea. They managed to show that the Bohr quantization condition followed from Heisenberg's assumptions. They erased most of Heisenberg's worries but introduced a new one when they found that the magnitude of the non-commutation of the product of the matrices Q (position) and P (momentum) was imaginary, that is, it involved the square root of minus one, specifically.

$$QP - PQ = ih/(2\pi)$$

where the quantity $i = \sqrt{(-1)}$. The occurrence of an imaginary quantity as a primary ingredient in a physical theory had never been seen before and was potentially more worrying than the appearance of matrices with their noncommutation. The concern did not last long, however.

In England, Paul Dirac, a young graduate student who had contemplated a career in electrical engineering, but had ended up in theoretical physics, showed that the strange mathematics in Heisenberg's theory was essentially, in his words, a "Hamiltonian dynamics of matrices". He showed that Heisenberg's quantum theory was a natural extension of the elegant formulation of Newtonian mechanics written by the Irish physicist William Rowan Hamilton in the nineteenth century. The success of the theory was assured when Wolfgang Pauli managed to compute the emission spectrum of hydrogen directly from Heisenberg's theory.

The creative work in this enterprise had all been done by men in their early twenties (Heisenberg, Jordan, Dirac, and Pauli), with guidance coming from an older man, Max Born, then in his early forties. It was (and is) widely believed, with some justification, that the innocent view, energy, and courage of youth are necessary in order to make the bold leaps required for progress in physics. This prejudice holds even more strongly in mathematics; remember that Werner Heisenberg had been discouraged at age 19 from going into mathematics. Well, exceptions do happen. The next player in this little drama was not, by any measure, a young man.

Erwin Schrödinger was born in Vienna in 1887. He showed an early talent for mathematics and entered the University of Vienna in 1906 to study physics. Ernst Mach, retired after his stroke, but still active and influential, did not have the positivist effect on Schrödinger that had taken so thoroughly on Heisenberg. Schrödinger was at the top of his class and great things were expected of him but this was not to be, at least not right away.

Schrödinger had very broad interests. In addition to physics, he involved himself with philosophy, Eastern mysticism, and had a keen eye for the ladies. He also wrote romantic poetry. In a word, he spread himself rather thin. After World War I, in which he was an artillery officer on the Italian front, he pursued an unexceptional career in theoretical physics with teaching positions at Jena, Stuttgart, Breslau, and Zürich. In 1924, Schrödinger read de Broglie's dissertation on matter waves and set to work to find an equation for the waves that would apply to a bound electron. By 1925, he had succeeded in writing down a relativistic equation and was devastated when the attempt to apply it to hydrogen failed. This failure depressed him for several months.

In the winter of 1926, Schrödinger gathered himself for another attempt on the problem. He packed his papers and he, together with his current mistress, set off for a rented chateau in the Swiss Alps. There he wrote a non-relativistic version of his earlier equation. Lo and behold, it worked! The relativistic equation that he had written in 1925 was later discovered by Klein and Gordon, whose name it now bears. The reason for the failure of Schrödinger's attempt to apply the Klein-Gordon equation to the hydrogenic electron is that the electron has a non-zero *spin*, a sort of internal quantum angular momentum. Spin is basically three-dimensional, whereas the Klein-Gordon equation, being relativistic, is basically four-dimensional. The Klein-Gordon equation works quite well for zero-spin particles. Later on (1930), Paul Dirac would succeed in writing a relativistic wave equation that is not only capable of dealing with spin, but specifically includes it in the equation.

Note that with Heisenberg, Born, and company, it was particles all the way. Schrödinger, on the other hand, was taken with de Broglie's idea and reasoned that matter might well be basically wave in nature. The Schrödinger equation has several advantages over the matrix mechanics of Heisenberg. The logic of the wave equation is much more straightforward than the matrix approach and is easier to apply in most cases.

In that first burst of creativity, Erwin Schrödinger, long past the creative age (he was 38), completely solved the hydrogen atom with a little help from his friend, the mathematician Hermann Weyl. This led to something of a crisis in the physics community. How could Heisenberg's matrix mechanics and Schrödinger's wave mechanics, which resembled each other not at all, both give the same answers? Schrödinger himself sorted this question out. He managed to establish the complete mathematical equivalence of the two theories.

Schrödinger's theory, up to this point, was time-independent. He showed that the usual time-dependent wave equation that worked for light would not work for electrons, and came up with a different time-dependent equation that would. He was disturbed to see that the time-dependent term contained the imaginary factor i (square root of minus one) and was not at all sure what that might mean. The equation had one overwhelming virtue that dispelled any gloom cast by the factor i. The equation was linear, and hence easily solvable. Because of this linearity, a sum of solutions was also a solution and so special solutions could be easily constructed by Fourier analysis (the addition of sine waves of various frequencies and amplitudes). The question remained, however, what was it that was waving? What was the physical significance of the wave function, which might well be complex, that is, composed of real and imaginary parts? Schrödinger claimed that the square of the wave function (which is real) was some sort of weighting factor, coming close to what would later be shown to be the truth. In the next breath, he stated his belief in the reality of the matter wave (he was a realist, not a positivist). In this way, he turned from the very brink of a potential conceptual breakthrough and left this critical step for someone else, another older man.

In 1927, Schrödinger, at the invitation of Max Planck, accepted the chair of theoretical physics at the University of Berlin. There he formed a close friendship with Planck and had six productive years. In 1933, troubled by the excesses of the Nazi regime, Schrödinger left Germany. After batting around Europe in search of stability, he arrived in Ireland with his wife (and mistress) and settled in at the Dublin Institute for Advanced Studies for a 17 year stay. There he worked on problems in general relativity, cosmology, and quantum biology. The latter work resulted in a short and very interesting little book entitled, *What is Life?*. In 1956, he returned home to Austria and died there in 1961.

The older man (older than Schrödinger by five years) who solved the riddle of the wave function was Max Born. He correctly identified the square of the complex wave function $\left| \psi^2(x.t) \right|$ as the *probability* of finding the electron at the location x and at the time t. Thus the wave function ψ, which contains all that we can know about the electron, is the complex (and hence not real in the mathematical or physical sense) amplitude, whose square (the intensity of the wave) is a probability. This implies that all that we can know about basic matter is probabilities. This was a far cry from Schrödinger's notion of a real physical wave underlying a real physical electron and he was not at all happy with it. *Born had effectively destroyed determinism*, nothing less. Determinism, the belief in an unbroken mechanistic sequence of cause and effect, had not really been in good health for quite some time, what with wave-particle duality, but most people hoped that determinism might recover its health. Although it would take a long time to accept it (most people have yet to do so), determinism had been done to death in no uncertain terms and the world was just going to have to deal with that. Most physicists dealt with it by taking a pragmatic view (calculate, but don't think). Logical positivism acquired new adherents. Many of the older generation such as Einstein were saddened and confused by the way things were going. Einstein was sure that: "God does not play dice with the World." "Don't tell God what to do," retorted Niels Bohr, who was the reigning high priest of the quantum religion by virtue of his seniority and the force of his personality.

This chapter began with Werner Heisenberg and it is going to end with him. In 1927, he was in Copenhagen. Bohr was on vacation and Heisenberg was working alone. The mathematicians had been overjoyed that one of their conceptual inventions, matrices, had been found to have application in physics. The physicists, on the other hand, had not taken all that well to a theory containing strange and unfamiliar mathematics, even if it did give the right answers. Schrödinger's equation, by contrast, was a linear differential equation, something with which physicists were thoroughly comfortable. Werner Heisenberg was struggling to attach some physical meaning to his matrices.

He realized that all of the concepts in classical physics have their counterparts in quantum theory. Momentum, position, energy, time - all have the same meaning in the microworld as they do in the macroworld. As we will see, however, when you try to pin down one attribute, another gets away from you entirely, thereby creating an irreducible *uncertainty*. The key was in looking at pairs of attributes, specifically those whose matrix multiplication does not commute such as x, p_x (position and momentum) and t, E (time and energy). Heisenberg devised a thought experiment now known as the *Heisenberg microscope* to explain the concept. This is shown in Fig. 7-1.

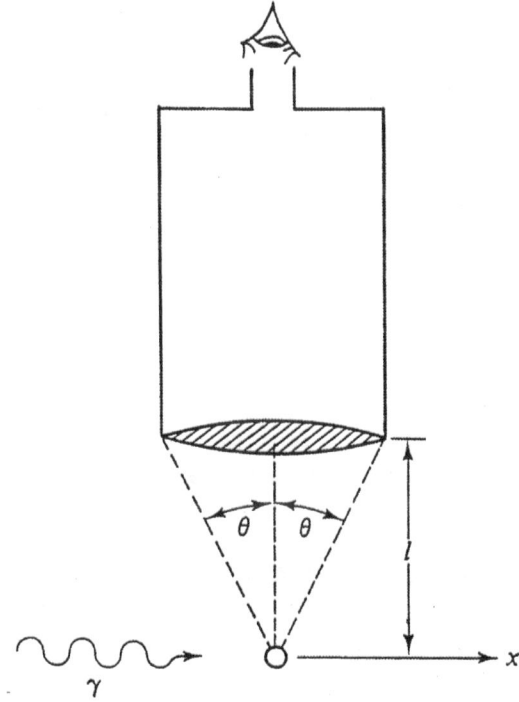

Figure 7-1 The Heisenberg microscope. An electron is viewed through a microscope of aperature diameter D at distance ℓ from the object lens. It is illuminated side-on by a single photon moving in the x-direction.

A particle, an electron, say, is placed under the lens of a microscope of given aperture diameter D and focal length ℓ. In order to observe the electron, we must illuminate it. For this purpose, we will employ a single photon incident onto the electron from the x-direction at right angles to the axis of the microscope. Any optical instrument is ultimately limited in its resolving power by diffraction effects. The angular resolution δφ is proportional to the wavelength λ of the light used for illumination and inversely proportional to the diameter D of the object lens. The positional error of the electron owing to these optical limitations is therefore on the order of.

$$\Delta x \approx \ell \lambda / D.$$

Since we are obviously limited in practice as to how large we can make the aperture diameter D, our choice will be to reduce the wavelength λ in order to reduce the positional error.

When the light, limited to a single photon (but not less), strikes the electron and is scattered into the reception cone of the microscope, a certain amount of momentum is transferred to the electron. A photon scattered anywhere within this cone will be transmitted to the eyepiece. Within these angular limits, the momentum in the x-direction lost by the photon and hence gained by the electron, can be shown to be uncertain by an amount.

$$\Delta p_x = p \sin\theta = (hD)/(\lambda\ell)$$

Where Planck's constant h enters via the de Broglie relation, $p = h/\lambda$. As an aside, I should note that this microscope thought experiment of Heisenberg's is somewhat ironic in light of the fact that he failed to know the resolving power of such an instrument on his Ph.D. oral examination.

If we now take the product of Δx and Δp_x, we find that the product of the uncertainties is.

$$(\Delta x)(\Delta p_x) \approx (\ell\lambda/D)(h/\lambda)(D/\ell) = h.$$

The implication here is that if you try to measure the position of an electron with a precision Δx, the act of measurement will unavoidably impart a momentum error Δp_x such that the product of the errors is of the order of Planck's constant h. This *disturbance* interpretation of what Heisenberg called the principle of *unbestimmtheit* (indefiniteness), what we now call the *Uncertainty Principle*, is true as far as it goes, but it does not capture the basic meaning. The uncertainty in the observation of the electron does not arise owing to the irreducible energy of a photon and the delicacy of the electron, as though the electron had a definite position and momentum before we disturbed it. This is the classical mindset at work. The origin of the uncertainty principle is more subtle than that. It is bound up in the dual wave-particle nature of the electron. In order to explain why this is so, I need to subject you to a small bit of mathematical reasoning. I hope that you will bear with me. I shall make the process as painless as possible.

Fourier analysis, as we have mentioned before, is the process of representing an arbitrary function as the sum of sine and cosine waves of various amplitudes and wavelengths. A Moog synthesizer, for example, is essentially a Fourier analyzer. It generates a variety of pure sinusoidal tones and superimposes them to duplicate the sound of any instrument in the orchestra. The waves add together in such a way as to reinforce where the amplitude of the function to be analyzed is large and to cancel where the amplitude is small. If the function in question is broad, the range of wavenumbers ($k = 2\pi/\lambda$) required is modest. If the function is highly localized, such as the wave packet that we showed in Fig. 6-7, then a considerable range of wavenumbers is required. It can be shown that for a Gaussian (bell curve) wave function, $\Delta k \Delta x = \frac{1}{2}$, and for any other function

$$\Delta k \Delta x \geq \frac{1}{2}.$$

If we now take the de Broglie relation in the form

$$p = h/\lambda = hk/(2\pi)$$

into account, we find that

$$(\Delta x)(\Delta p_x) \geq h/(4\pi).$$

This is Heisenberg's uncertainty principle in its most general form.

Defining the angular frequency ω, which is related to the circular frequency v by $\omega = 2\pi v$, it can be shown that the uncertainties in the product of angular frequency and time must obey

$$\Delta \omega \Delta t \geq \frac{1}{2}.$$

With the Planck relation, written in the form

$$E = hv = h\omega/(2\pi)$$

We find for the energy-time uncertainty relation

$$(\Delta E)(\Delta t) \geq h/(4\pi)$$

All of this has nothing to do with the *disturbance* of some pristine microscopic deterministic state at all. It is not that some precise momentum and position are destroyed in the act of measurement. In fact, *neither attribute exists until the act of observation creates it*. We will have a good deal more to say in regard to that startling statement. In the meanwhile, we can summarize Heisenberg's uncertainty principle as follows.

The wave-like properties of an electron require that it be spread out over a finite volume. In other words, the electron *must*, at one and the same time, occupy a range of positions. The particle-like properties require that the electron must be localized, and in order that it still retain its wave-like nature, it must simultaneously possess a range of momentum values. Heisenberg's uncertainty principle states that the range of positions is inversely proportional to the range of momenta.

The confirmed predictions of this principle are not limited to statements about the maximum accuracy with which simultaneous measurements of position and momentum or energy and time can be made. It can be shown, for example, that in highly compact objects such as white dwarf stars where matter is squeezed so tightly together that the electrons are highly constrained in their movements, the electron momentum increases and a pressure arises owing to this quantum mechanical effect. This is known as *degeneracy pressure* and is independent of temperature. It is this degeneracy pressure that supports the star against collapsing under its own gravity. Black holes evaporate and the vacuum seethes with virtual particles winking in and out of existence, all because of the truth of Heisenberg's principle.

Let us not loose sight of the forest for the trees. The most important result of this chapter by a country mile is that determinism, following a long illness, was finally euthanized by Dr. Born. Make no mistake. At the most basic level of existence, determinism is gone and will not be back.

CHAPTER EIGHT

AND YOU THOUGHT YOU HAD HEARD THE WORST

in which an analysis of measurement
consigns objectivity and locality to the
boneyard along with determinism.

How do we get from the wave function and its undulating mist of probabilities and potentialities to the world of appearances where everything abides in no more than one place at a time and either is or is not? Simple, we make a measurement. If you are a pragmatist, that answer probably satisfies you. But you are probably not a pragmatist, otherwise you would not have read past the Preface.

From the point of view of the wave function (and in quantum mechanics, *everything* is from the point of view of the wave function), we note that in the course of the measurement, the wave function must change discontinuously. At one instant the wave function still contains a range of probabilities describing possible outcomes of the measurement. In the next instant, the measurement is accomplished and the probability amplitudes of all but one of the potentialities will have collapsed to zero. The probability amplitude of the remaining potentiality (now a certainty) will be unity. The interesting thing to note is that the Schrödinger equation only describes smooth continuous changes of the wave function, so the answer to our question of how the collapse occurs is not contained within the theory. This makes for an interesting situation, to say the least. We are on our own with this one and will have to find clues where we can.

Let us initiate our quest by asking whether or not a classical measuring apparatus can collapse the wave function, that is, promote a potentiality to actuality status. It would seem that the answer must be no. All material objects are described by quantum mechanics. We will examine this statement in depth in the next chapter, but please accept it for now. In the classical limit of the world of appearances, the *graininess* of the quantum world appears smooth and continuous owing to the very small value of Planck's constant (h = 6.63 x 10^{-34} Joule second). If you insist that classical measuring devices (photomultipliers, Geiger counters, and the like) are somehow nonquantum objects and are capable, by some means, of collapsing the wave function, then you are going to have to come up with an argument to substantiate your claim. Is there some sort of boundary between objects containing fewer than some magic number of electrons and those containing more that makes the first a quantum object and the second a classical object, somehow empowered to make quantum decisions? Put that way, I think very few people would support the notion.

Suppose now that you have a chain of measurement devices. For example, you might have a Geiger counter ready to trigger an electrical pulse in the event of a nuclear decay (a random quantum event). The output from the Geiger counter is wired into the input of an oscilloscope whose cathode ray tube will be photographed by a Poloroid camera. You can, in this way, set up a deterministic measurement chain, but the last link, in this case, the image on the film, will remain as a coherent superposition of quantum probabilities until you look at it and see whether the decay occurred or not.

This is the point (and it is a Hell of a point). We know that a human observer can collapse the wave function. We have no evidence at all to suggest that a nonsentient material device can do it. We do know that all material devices of whatever size are subject to the laws of quantum mechanics. Then, finally, we know that this supremely successful theory contains no mechanism for the collapse of the wave function.

Is it the act of observation or perception that does the job? I think not. Mechanical devices can perceive and record their perceptions. It is the mind and, specifically, consciousness that must register the event and promote it to full reality status in the world of appearances. All material agents (including the brain) are subject to the laws of quantum mechanics. Only a non-material agent will do the job. So we are forced to the conclusion that consciousness and the mind are not simply epiphenomena of the brain and that it this transcendent aspect of ourselves that is responsible for actualizing quantum phenomena in the measurement process.

As Sherlock Holmes famously noted, "When all alternative explanations have been eliminated, whatever remains, no matter how unlikely, must be the truth."

We have clearly strayed out of the domain of physics and into the realm of philosophy (the mind-body problem), psychology (mind and consciousness), and quite possibly, religion. There is no reason in the world not to do so. If we examine the areas of knowledge that each of these activities staked out for itself, we find a great deal of overlap. The quantum questions cannot be pursued without taking such a multidisciplinary approach.

Which of us are empowered to make quantum decisions? Must one have a Ph.D? Eugene P. Wigner, the Hungarian physicist and philosopher described a puzzle that has come to be called *Wigner's friend*. Wigner sends his friend to check on the results of a quantum measurement. Until the friend returns and tells him, Wigner is unaware of the outcome. Does this imply that the wave function, now including the friend, remains uncollapsed until Wigner is made aware? Does the friend exist in a coherent superposition of states until his act of telling Wigner collapses the wave function? Certainly not. You might imagine Wigner sending one person, who sends another, who sends another, ad infinitum. Is only Wigner's awareness capable of collapsing the wave function? That is the same, as near as not, as suggesting that only Eugene Wigner exists. That notion, of course, is *solipsism*, which, as Schopenhauer commented, needs not so much a refutation as a cure.

Let us suppose then that man has a mind-consciousness complex that is basically immaterial and not merely an epiphemomenon of his material brain and that this mind-consciousness is the agent responsible for collapsing the wave function. In that case, we can hardly continue to regard ourselves as subjective independent observers of an objective world that is entirely external to ourselves. Heisenberg's uncertainty principle, which we discussed in the last chapter, had already cast a shadow on the notion of an external objective world. Our conclusion that the critical step from potentiality to actuality is accomplished by the mind-consciousness complex certainly implies that objectivity must be laid to rest along with determinism.

I have used the term mind-consciousness complex up to this point not to demonstrate my expertise in psychology, but rather to hide my ignorance. We will devote an entire chapter each to the psychological, philosophical, and religious implications of quantum mechanics. We have uncovered so many fascinating avenues of investigation that it is tempting to rush down all of them at once. I will attempt to restrain myself, however, and will stay with the physics until we have learned all that it has to teach us.

Another question that might well be asked is why one potentiality comes to be actualized and not another. This, I think, is the question that most troubled Einstein and evoked his comment about God not rolling dice. This question was answered in 1957 by Hugh Everett III, at that time a graduate student of John Wheeler's at Princeton. Everett proposed that no choice is made but that all possibilities occur. According to this idea, the universe splits into non-communicating branches, in each of which a different quantum probability is actualized. It is a measure of the desperate conceptual straits that the analysis of quantum measurements had engendered that this *Many-Worlds* proposal found favor with quite a few people and still does. Note that Everett's theory does not explain quantum measurement in general. You are still stuck with the mind as the actualizing agent. Many-Worlds does not raise objectivity from the grave. It only answers the question of how the choices are made by suggesting that choices are not made at all. According to Everett, every time we carry out an experiment in which we measure an electron with spin up, the universe creates a near copy of itself in which we measured spin down. It's probably fair to say that the people who accept the Many-Worlds idea do not love it. They simply hate it somewhat less than they hate not having an explanation, even such an untidy one. Make no mistake. For gross extravagance, the Many-Worlds theory is unbeatable. At every instant, a zillion times a second, the mass and energy of the universe must be doubled. To my mind, this is a very high price to pay for God to keep his dice in his pocket. We will return to this question of how quantum decisions are made later in the book and will be able to argue strongly that Everett's proposal is not the answer.

A discussion of the quantum measurement problem would not be complete without a mention of *Schrödinger's Cat*. Erwin Schrödinger, as you may remember, was something of a naive realist. The idea that unobserved phenomena exist only as a superposition of probabilities was preposterous to him and in 1935, he introduced the following thought experiment to drive home his point.

A cat is sealed in a room with no windows. The room also contains a *demonic device* consisting of a radioactive isotope, a Geiger counter, and a mechanism that will release deadly cyanide gas to kill the cat if the Geiger counter is triggered by a decay event. For the hour that the cat is to be sealed in the room, the probability of a decay event is exactly fifty percent. We will not know until we open the room at the end of the hour (with our gas masks on if we know what is good for us) whether we will find the cat alive or dead. The interpretation of quantum mechanics that we have argued in this chapter, which is, in fact, the conventional one, would have us believe that the cat is existing for that hour in a state of coherent superposition, half alive, and half dead. This notion of a half alive, half dead cat was so unreasonably bizarre to Schrödinger that he assumed it would stop the observer-created reality interpretation in its tracks. This is worth a bit of discussion.

If Schrödinger had arranged for a hammer to be triggered to break a vase, the impact of his thought experiment would not have been as provocative as it was. Clearly the unobserved radioactive isotope is in a superposition of states (decayed and undecayed) and this superposition is linked unambiguously to the hammer poised to break the vase. So the whole isotope, Geiger counter, hammer, vase system is included within the wave function. No one will object to that. It is the fact that we have a living sentient creature at the end of the chain that puts the fox amongst the chickens. Let's go to the extreme of supposing that we had sealed a living human being, say Erwin Schrödinger, in the room. Would he have had to be regarded as existing in a half live, half dead state for the hour that he is out of communication with us? Now we are back to a version of the Wigner's friend question. Suppose everyone in the world except for you and me were sealed in that room. Would they all exist in a zombie state waiting for us to collapse their wave function? Of course not! Solipsism does not obtain. Schrödinger and any number of others sealed into the room with him are perfectly capable of actualizing their own situation.

So humans never find themselves in a half alive, half dead state, whereas an unobserved vase may exist half smashed, half whole. What about a cat? Is a cat sufficiently far up the ladder of evolution to be able to promote quantum events into the world of appearances? That is a very interesting question. It is not a physics question and it is certainly not the question to which Schrödinger was trying to call attention. I would say that this question is properly in the domain of religion. Western religions (Christianity, Judaism, and Islam) focus their entire attention on human beings in their relationship to God. In the esoteric Eastern religions of Buddhism and Hinduism, the focus is on all sentient beings, who may incarnate in the animal realm as well as the human realm and who have within themselves great (mostly unrealized) spiritual powers. My answer to whether or not a cat has the power to collapse the wave function will have to take the form of a definite maybe.

Those people who, along with Schrödinger, find the idea of a half alive, half dead cat too preposterous to endure, often take refuge in the statistical ensemble interpretation of quantum mechanics (which I was taught in school as gospel). This interpretation maintains that quantum mechanics only applies to large statistical ensembles of identically prepared measurements, for example an astronomical number of cats in an astronomical number of rooms. We can then say that approximately half of them are dead and the other half are alive, which preserves objectivity. Observation will back us up and we need not deal with those uncomfortable coherent superpositions applied to single objects. All well and good, but nowhere is there a rule that says that quantum mechanics does not describe single objects or events. Quantum mechanics was, in fact, formulated to describe single events, so that large ensemble security blanket won't save you from the boogie man.

The logical positivist would say, echoing Wittgenstein, that the questions that we have been asking are not answerable and should never have been asked in the first place. In the present context, this position amounts to sophistry. Sorry, but it does.

We have revealed a number of bizarre features of quantum mechanics, features that must cause us to abandon determinism (the unambiguous evolution of the world as some great clockwork machine) and objectivity (the notion that we can regard the world as the entirely independent object of our observation). It is ironic that an experiment first carried out by Thomas Young in 1803, the experiment that established the wave nature of light, is also the experiment that displays the entire range of quantum weirdness.

In this experiment, Young caused sunlight admitted through a pinhole to shine on a screen containing two closely spaced holes. The light from these two holes was shone on a wall. If either of the holes in the screen was covered, then the resulting pattern on the wall was a more or less spread out bright image of the hole. If both holes were open, the pattern on the wall was not the simple sum of contributions from the two holes considered singly. It was an alternating pattern of light and dark with the central light bands brighter than the ones on the wings. This is shown in Fig. 8-1, which also shows that when the path difference from the two holes to the imaging screen is an odd integral multiple of half-wavelengths, then a dark band results. There the two waves arrive 180 degrees out of phase and interfere destructively. When the path difference is an integral number of whole wavelengths, then the waves from the two holes are in phase and the interference is constructive. By measuring the spacing between the fringes, Young was able to calculate the wavelength of the light. The experiment made a convincing case for the wave nature of light, which, as we have seen, was confirmed theoretically by Maxwell about 60 years later.

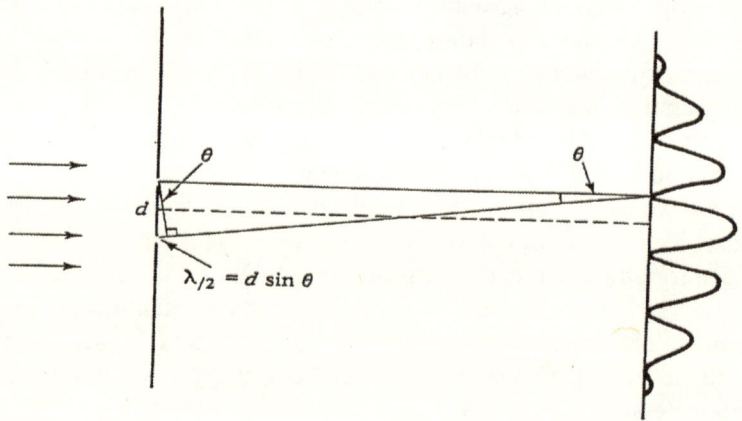

Figure 8-1 The Young two-slit experiment. The interference pattern at the right shows the alternate constructive and destructive superposition of coherent waves emanating from the two slits.

With the advent of quantum mechanics, we now know that light is emitted in irreducible bundles of energy that we call photons and is detected as photons, if the light is dim enough that individual photons can be distinguished. This is technically feasible. The obvious question is, will the interference effect persist using light so dim that only one photon is in the experiment at a time? In other words, can one photon simultaneously pass through both holes and interfere with itself? In the classical world of appearances, such a thing would be unthinkable. The photon, regarded as a classical object could not be in two places at once. Quantum mechanics predicted interference. The experiment was carried out by a team in Paris in the 1980s. Quantum mechanics was right. The interference pattern could be seen to build up, dot by dot, as each photon arrived at the detection screen. The pattern was not obvious initially, but it soon became apparent that the photons were registered only in those regions that would build into bright fringes. Could Young's experiment also be carried out successfully using the de Broglie waves associated with electrons? A Japanese group in Tokyo showed that this could be done in 1987. In Fig. 8-2, we see the build-up of the two-slit interference pattern using electrons. The result for photons is identical. The pattern has also been produced using atoms. In our minds, electrons and atoms are more *thinglike* than light, and so the question of how a single atom or electron manages to be at two places at once is even more jarring.

Perhaps we can sneak up on the electron and see how it does this trick. We close one hole and the interference pattern vanishes. We simply build up a bright spot behind the open hole. No surprises there. We can shine a light on the open hole and observe light scattering by the electron as it passes through. Nothing there to upset a classical physicist. Now let's open both holes and monitor one of them with our light beam to see through which hole the electron passes. When we do that, we may detect the passage of an electron through the monitored hole, or we may not, showing that it went through the other hole. In *either* case, there is no interference pattern. How does an electron passing through the right hole know that the left hole is being monitored? It is as though the electrons have some sort of *non-local awareness*, something that Einstein (in another context) referred to as "spooky action at a distance." Set up the particle detector on one slit, but don't turn it on, and the interference pattern appears. Switch the detector on, and no matter whether it detects an electron to tell us that the electron went through the monitored hole, or not, implying passage through the other hole, the interference pattern is no longer present in either case.

The quantum mechanical explanation for this remarkable state of affairs can be understood in terms of the uncertainty principle. When we locate the electron (or photon) by determining through which slit it passes, we increase the momentum error and consequently the range of wavenumbers. This wipes out the interference, which requires a small range of wavenumbers. The interesting point is that the entire experiment including the experimenter and his knowledge, direct or indirect, is required to predict the outcome. Determinism and objectivity are both shown to be inapplicable to this experiment. But wait, even stranger things are to come.

John Wheeler suggested that it might be interesting to place the electron detectors not at the slits, but between the slits and the viewing screen. In that way we might be able to fool the electron into revealing itself as a particle or wave after it has passed the slits. Quantum theory says that if we gain knowledge of the electron in one or the other channel, directly or indirectly, when it passes the slits *or after*, there will be no interference. If we don't turn the detector on, then interference will return. The startling thing here is that we are determining the behavior of the electron at the slits by an action that takes place after the electron has already passed them. The *delayed choice experiment*, which was a thought

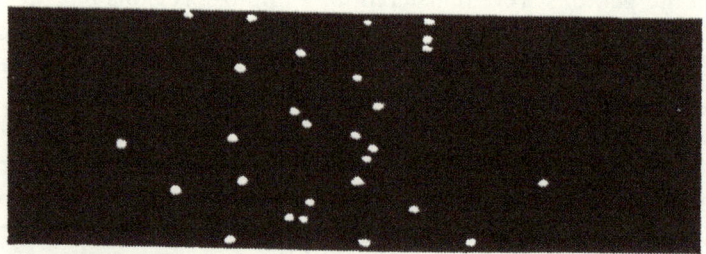

(a) After 28 electrons

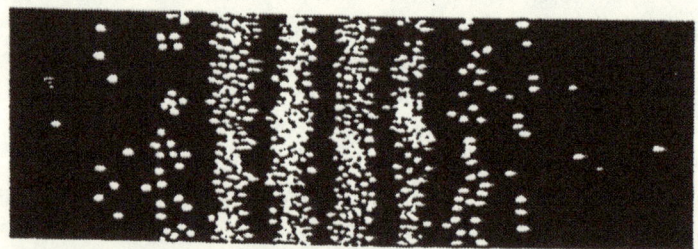

(b) After 1000 electrons

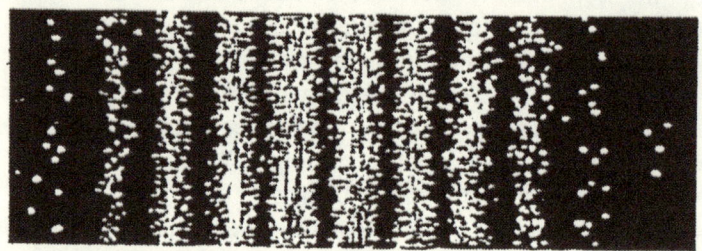

(c) After 10,000 electrons

(d) Two slit electron pattern

Figure 8-2 The two-slit interference pattern produced by successively increasing numbers of electrons.

experiment when Wheeler proposed it, can be further refined by using very fast switches to turn the detector on only after the electron has passed the slits. This arrangement will force the electron to have non-local awareness and a knowledge of whether or not the detector will be switched on in the future, in other words, a modest precognition.

The experiment was tried in the 1980s by two groups, one at the University of Maryland and the other at the University of Munich in Germany. It worked as predicted. The electron could not be fooled by the delayed choice feature. A reversal of the arrow of time defined by the normal cause-effect sequence must now be added to the confirmed list of freaks in the quantum zoo.

Wheeler also proposed a cosmic version of the delayed choice experiment. In the chapter on general relativity we mentioned that light from a distant star or quasar can be bent around a galaxy or other massive object that lies on the path between the object and Earth. If the alignment is rather precise and the intervening object is symmetric, light from the distant object will curve uniformly around and we will see it as an Einstein Ring. If the alignment is not so perfect, as is usually the case, then multiple images (generally two) will be formed. One such Einstein Ring was found in 1987. We are more interested in the double image case, however.

Using conventional optics, we can steer light from the two images in such a way as to cause the beams to cross. The light is dim enough that individual photons can be delineated. If we choose to place our photo-sensitive screen at the crossing point where the two beams are superimposed, then the whole setup becomes a giant Young two-slit experiment and we should expect to see interference fringes. In effect, our choice of monitoring the beam intersection here and now will have told the photon to travel around *both* sides of the intervening galaxy several billion years ago. If we also choose to monitor one or the other individual beam, then we will destroy the interference pattern and will, in effect, have told the photon to travel around one side of the lensing galaxy. This could hardly be described as modest precognition. We are, in effect, reaching back to a time before our Sun had begun to shine.

The cosmic version of the delayed choice experiment is still a thought experiment, but its results cannot be doubted. The difference between this proposal and the experiments carried out at College Park, Maryland and Munich are only quantitative and not qualitative. The problem of carrying out the cosmic experiment is one of *coherence time*. If we separate light from one source into two beams and send the beams along paths of different lengths so that they arrive back at a common focal point having quite different ages, then the coherence will have been lost and the two beams will no longer be able to interfere. It is rather like having two groups of soldiers marching in step with one another. March the two groups to opposite sides of town and then bring them back together. They will no longer be in step. That's the way it is with the light. In traveling around the two sides of a galaxy several hundred thousand light-years across, the light paths are going to be several light-weeks different at the very least and that time is much greater than the coherence time of light from the distant quasar.

In 1993, a much more compact gravitational lens was observed. An unseen gravitational source in our own galaxy (probably a neutron star or a black hole) passed in front of a dim star located in another galaxy causing a succession of lensed images of the distant object to pass across the line of view. If the intervening body that does the lensing is no more than planet size, then the path differences should be well within the coherence time of the light and someone should be able to slip a rapidly switched detector into one of the beams and erase the interference pattern.

Thomas Young's experiment of 1803 lives on in newer and stranger incarnations. This one experiment shows us that the world is both indeterminate and non-objective. Decisions that we make now can give rise *instantaneously* to effects throughout space and time. We must, therefore, add *non-locality* to our list of properties of the reality that underlies our local, determinate, objective world of appearances.

CHAPTER NINE

THE GOSPEL ACCORDING TO ST. NIELS

In which we begin to suspect that there
is something rotten in the state of
Denmark.

Not only is quantum reality vastly different from classical reality, quantum mechanics is not even a theory in the same sense as classical theories. In the first place, there is no statement of principles. Classical theories, even when they are statistical, always have deterministic, objective underpinnings, and thus it is possible, with our experience of the world, to have some notion of 'how the little men are running around down there', some mechanistic mental image, some intuitive feel for the thing. Not so with quantum mechanics. Successful as it was and is, quantum mechanics needs to be augmented by an *interpretation*, some words of comfort from a figure of authority to tell everyone that things are all right and that the ground beneath our feet is not about to vanish into a haze of inconceivable quantum probabilities. The man who, more than anyone else, filled that role by virtue of his qualifications and the strength of his personality was Niels Bohr.

Bohr had been given his own think tank in Copenhagen, courtesy of the Carlsberg Brewery, and had, as we have seen, attracted many of the young men whose courage and innocence of mind led to the development of quantum mechanics. Bohr was used to being the older man, providing guidance. He enjoyed the role and continually expanded upon it. The rationale that he and his colleagues came up with for quantum mechanics quickly came to be called the *Copenhagen Interpretation*. It was generally accepted by the physics community from about1929 until some time in the mid 1980s. There were some holdouts, of whom the most prestigious was Albert Einstein. He began a series of debates with Bohr, first trying to find an error in the uncertainty principle, and later questioning the completeness of quantum mechanics. Einstein gave Bohr a few sleepless nights with the clever thought experiments that he produced, but Bohr was always able to show how quantum mechanics could answer the challenge. Bohr's statue as the high priest of quantum mechanics grew and Einstein became more and more a peripheral figure in physics.

Let me interject an observation here before we go on to discuss the Copenhagen interpretation. As a dogmatic classical physicist, Einstein was wrong about quantum mechanics, that is to say, his concepts were wrong. His instincts, operating beneath the conscious level, were generally right on the money, however. He had an unerring knack for *asking the right questions*. He usually came up with the wrong answers where quantum mechanics was concerned, but that is not so important, in retrospect.

The security blanket that we refer to as the Copenhagen interpretation included the uncertainty principle and attempted to clarify the wave-particle duality issue, the nature of measurement, and the range of applicability of quantum mechanics.

The wave-particle duality issue was addressed in the *Principle of Complementarity*. In this principle, Bohr noted that in order to describe phenomena completely, one must sometimes use contrasting pictures. These pictures only appear in mutually exclusive situations. In other words, particle or wave, but never both at the same time. In this, Bohr insisted that classical reality (particles) be untainted by quantum reality (probability waves). At first thought, our description of the two-slit experiment of Young would appear to bear this out. Look for the presence of an electron or photon in either channel and the interference pattern disappears. But what happens as we turn down the intensity of the light we are shining on one slit, the light that is scattered by the electron, thus revealing its presence? Does the detector light have to be completely extinguished in order to reveal the interference pattern? No, it doesn't. As the intensity of the detector light is reduced, the interference pattern begins to appear. It would seem that there is a patch of grey between black and white, a small zone in which you see both particle and wave.

If you stop and think about it, the mingling of wave and particle was before our eyes all along. In the Planck relation, $E = h\nu$, we are told that the energy of the particle aspect is proportional to the frequency of the wave aspect. In the de Broglie relation, $p = hk/(2\pi)$, we are told that the momentum of the particle aspect is proportional to the wavenumber of the wave aspect. This would seem to me to constitute an intermingling of particle and wave.

A more decisive experiment so far as the complementarity question is concerned was carried out in 1992 by Yutaka Mizobuchi and Yoshiyuki Ohtake of Hamamatsu Photonics in Japan. The experiment was conceived by Dipankar Home and Partha Ghose of the Bose Institute in Calcutta and Girish Agarwal of the University of Hyderabad. The idea was to try and force wave and particle both to appear simultaneously. A pair of right-triangular prisms are arranged with their hypotenuse sides separated by the smallest imaginable gap, amounting to about a tenth of the wavelength of the light to be used. A single photon light source emits a photon perpendicular to one of the faces so that it enters the prism and encounters the hypotenuse face at a 45 degree angle. If there was only the one prism, the photon would be reflected off the inner hypotenuse surface and would come out of the other face at right angles to its initial path as shown in Fig. 9-1(a). If both prisms are present and are in solid contact with each other to make a square block as shown in Fig 9-1(b), then the photon will go straight through the block without being reflected at all. If, however, both prisms are present and are separated by a gap smaller than the wavelength of the light, then some of the light will be reflected and some will *tunnel* across the gap and continue in a straight line. The kicker is that only waves can tunnel across the gap, not particles. What happens, then, when you introduce a single photon into the system shown in Fig. 9-1(c)? The Japanese experiment was arranged so that the tunneling probability was exactly 50 percent. Detectors were placed in both channels. If these detectors only register at different times, this is proof that light is traveling through the experiment one photon at a time. In successive runs, half of the single photons were found in the reflection channel and half in the tunneling channel. The detector signals were not in coincidence, thus showing that the photons were behaving as particles at the gap and were not splitting and acting as waves by tunneling. The light thus behaves as photon and wave at one and the same time. Such single photon experiments were not possible in Bohr's time. It would seem that a major plank in the structure of the Copenhagen Interpretation has gone rotten. A flipped coin can occasionally land on its edge and show both faces.

The next part of the Copenhagen Interpretation that we shall examine is often not included in the interpretation. It is called the *Correspondence Principle*. Bohr invoked it as an argument in support of his angular momentum quantization relation in 1913. It has a bearing on another part of the Copenhagen interpretation, so it is appropriate to examine it now.

The correspondence principle states that the predictions of quantum mechanics and the predictions of classical physics agree in the limit of large *quantum numbers*. Quantum numbers are the integers that label the allowed orbits, angular momenta, energy levels, etc in a quantum object such as the hydrogen atom: n = 1 for the ground state, n = 2 for the first excited state, and so on.

It was often claimed that Bohr's quantization relation, which was the key to his 1913 model of the hydrogen atom was, in effect, an ad hoc assumption. Bohr even implied later on that such was the case. Bohr could be quite confusing and though people tended to hang on his every word, they were often not sure later what he had said. Bohr even confused himself at times. The fact was that Bohr had derived the angular momentum quantization using the correspondence principle, and a very clever bit of reasoning it was.

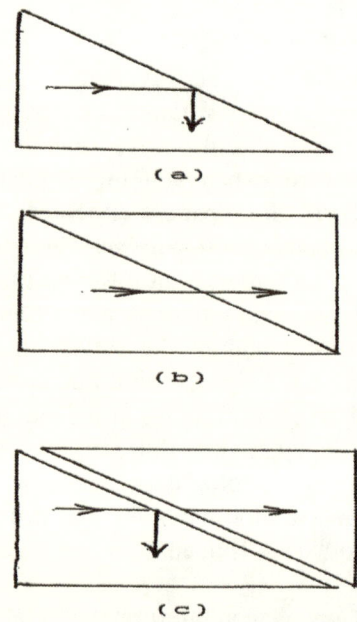

Figure 9-1 Hamamatsu beam-splitting experiment showing (a) internal reflection, (b) transmission, and (c) reflection and tunneling.

The correspondence principle states that quantum mechanics goes over into classical physics in the limit of large quantum numbers. Even though the correspondence principle had its origins in Bohr's 1913 theory of hydrogen, it applies equally to the Schrödinger theory. If the quantum number is sufficiently large, then the spacing between successive energy levels or successive allowed angular momenta shrinks to zero and the discontinuities that characterize the quantum world go over into the continuous changes that we expect in the world of appearances.

All this talk about very large quantum numbers is somewhat artificial. After all, we never see a hydrogen atom with an electron orbit having a radius of one centimeter (corresponding to a quantum number n ≈ 13,750), so why talk about it. A much more interesting question is, does quantum mechanics apply to the large collections of atoms with more normal energy levels that make up macroscopic objects? The answer is that yes, it does. We can show using the Schrödinger equation that the *expectation values* of position, momentum, and energy operators bear the same relation to each other as do the variables in Newton's equations of mechanics. Still, the question generates diverse opinions, but the arguments against lack credible scientific reasoning and usually take the form of wishful thinking designed to reassure the person that somehow he still lives in a deterministic, objective world.

Science invariably proceeds by evolution, not revolution. Newton's mechanics and Maxwell's electromagnetics coexisted uneasily at their point of overlap. Einstein's relativity healed the overlap by exchanging absolute space and time and a relative velocity of light for relative space and time and an absolute speed of light. In so doing, he established a theory that agreed with Newton's theory in the limit of low velocities, and correctly predicted phenomena at speeds approaching the speed of light. Special relativity did not replace Newtonian mechanics. It enfolded and extended it. If we wish to calculate, say, the dynamics of a billiard ball or the orbit of a satellite, we will certainly use Newtonian mechanics rather than the equations of special relativity. It's easier and in most cases, sufficiently accurate.

Similarly, general relativity extended and enfolded special relativity and Newtonian gravity, enabling one to make calculations in accelerated frames of reference including gravitational fields without the patchwork of inertial pseudo-forces. Unless distances are cosmic or gravitational fields are very large, the older Newtonian theory still does an adequate job.

Does quantum mechanics enfold and extend Newtonian mechanics? Yes it does. This is not as intuitively obvious as in the above-cited cases because we do not ordinarily experience quantum weirdness on the macroscopic scale of kilograms and meters, or at least we do not realize that we do. This is because of the very small numerical value of Planck's constant that obscures quantization on the large scale. Another thing that makes our world seem ordinary is that the wave functions of unwatched macroscopic objects do not spread with anything like the same speed that the wave functions of microscopic objects do. It can be shown that the wave function of an unwatched electron, assuming it to be a point particle, spreads with an infinite velocity. An unwatched electron is literally everywhere. The wave function of a single atom spreads at a speed of a few hundred meters per second. The wave function that characterizes an unwatched virus spreads with a velocity of approximately a ten billionth of a meter per second. For larger and heavier objects, the velocity with which the wave function expands from *particleness* to *waveness* is even slower. This is why the world of appearances seems so ordinary, so unlike our description of the quantum world. Little bits of naked quantum reality, like blackbody radiation and the photoelectric effect do intrude, and, of course, we are knee deep in miracles of quantum technology such as lasers, computer chips, transistors, and superconductors. The list is endless. There are also some more startling quantum phenomena that intrude, but you have to go looking for them. More later on this. For now, let's get back to the interpretations of quantum mechanics, Bohr's and others.

Bohr tells us in the Copenhagen interpretation that quantum mechanics gives us as much detail as there is. Not everyone agrees with that. There are those that insist that there are *hidden variables,* inherently undetectable forces at work that preserve determinism. In this view, electrons are particles with definite positions and momenta, observed or unobserved. The quantum behavior that we observe, for example in the Young double-slit experiment, is said to be determined by the action of an undetectable pilot-wave field, the hidden variables. If this field could be observed, then quantum probability would turn into certainty.

Realize up front that we are not talking science here. A theory can only be said to be scientific if it is *testable*. The fact that hidden variables are hidden and inherently undetectable takes this idea out of the realm of science.

De Broglie introduced a hidden variable interpretation of quantum mechanics in 1925, but it was never widely accepted, philosophically comforting though it was. The force of Bohr's personality proved the deciding factor. Then, in 1932, John von Neumann killed the hidden variable idea with his great book, the quantum bible.

Von Neumann was born in Budapest in 1903 and showed the talent quite early that was to make him into one of the greatest mathematicians of the twentieth century. In 1930, Paul Dirac in England had published *The Principles of Quantum Mechanics* in which he had attempted to give quantum mechanics a mathematical framework, something that it had lacked up to that point. In 1932, von Neumann came out with his *Mathematical Foundations of Quantum Mechanics*, which was a monster tome, more comprehensive than Dirac's book and a good deal less readable. Consequently, it was often quoted as the ultimate source of quantum authority, but was probably read by few. In the book, von Neumann claimed to have proved that no hidden variable theory could be consistent with quantum mechanics. This was accepted as Holy Writ until John Bell, an Irish physicist of whom we will hear more, caught the error in 1964. In retrospect, the error was a surprisingly trivial one for the great John von Neumann to have made. Even geniuses (and von Neumann was certainly that) have off days.

David Bohm, an American physicist who had run afoul of Senator McCarthy and his House Un-American Activities Committee and moved to England, wrote a hidden variable theory in the 1950s that appeared to duplicate the predictions of quantum mechanics. He was either unaware of von Neumann's impossibility proof, or just forged ahead anyway. In Bohm's theory, the hidden variable, a pilot wave, is altered by any attempt to measure position and momentum simultaneously. The result is an instantaneous (non-local) change in the shape of the pilot wave such as to give the uncertainty principle results. It worked, but it was not a pretty thing in the mathematical sense. It failed the aesthetics test, which is very definitely a consideration in theoretical physics.

You will have noted a theme of *dualism* running throughout the Copenhagen interpretation. Particle or wave. Classical or quantum. In describing the quantum measurement process, Bohr was *triplistic*. A quantum measurement, according to Bohr, had to involve the quantum object upon which the measurement was being made, the measuring device that was used to make the measurement, and the observer whose observation was necessary to collapse the wave function and promote one of the probabilities to actuality status. Quantum objects were fully quantum and could not be known. The measuring device was totally ordinary, a classical macroscopic instrument. Bohr stopped short of describing the attribute of the observer as transcendent; he was vague about how that piece of the process might work. He also did not attempt to justify how the measuring device, made of (quantum) atoms could be purely classical in violation of his own correspondence principle. Bohr appeared to think that the classical world and the quantum world were equally fundamental. Henry Margenau, who was Professor of Philosophy as well as Physics at Yale University saw only the quantum world as fundamental, a defining condition underlying the world of appearances.

John von Neumann addressed these same questions. For him, all material objects of whatever size, microscopic or macroscopic, were quantum objects. The aspect of the experimenter that enabled him to collapse the wave function was consciousness, not mere observation. Von Neumann was not in the security blanket business at all. He believed what the mathematics told him and took a *monistic* view of the measurement process.

Bohr did not question the Einsteinian concept of locality, even though some of the predictions of quantum mechanics called for non-local correlations. Locality was first successfully challenged in the 1960s (theoretically) and the 1970s and 80s (experimentally). We will devote a chapter to the description of this important work. We have already seen that the delayed-choice experiment suggested by Wheeler and carried out in the 1980s showed as well that the world is, in some sense, non-local. Does this mean that relativity is wrong? No, it doesn't, as we will see.

For now, it is time to look at some aspects of quantum logic.

CHAPTER TEN

BOOLE GOES DOWN FOR THE COUNT

in which we must make the choice
between physical weirdness and
quantum weirdness.

You have read up to this point and have either found the book interesting or
dull, which implies that either you have read to this point and found the book
interesting, or you have read to this point and found the book dull. A coin rests
on the table showing either heads or tails, which implies that either a coin rests
on the table showing heads or a coin rests on the table showing tails. These
statements are *logically* identical, differing only in their subject matter. In the
world of appearances in which we live, the logic reflected in these examples
seems obvious. We could as well say P and (Q or R) implies (P and Q) or (P and
R). You can insert any subject matter you like for P, Q, and R and the first
statement will still imply the second.

The study of logic was begun in the West by Aristotle. Given the Indian
talent for philosophy and mathematics, I assume that logic was codified in India
at an even earlier date. The expression of the rules that govern our logic in terms
of algebra was the work of George Boole, a nineteenth century Irish
schoolmaster. Boole is to logic much as Euclid is to geometry. They both
codified important aspects of the world of common appearances. I say common
appearances because we have already seen that for very long distances and/or
very large gravitational fields, so-called natural geometry is no longer Euclidean
but is one of the curved geometries described by Lobachevski, Riemann, and
Gauss.

The *distributive* law of logic, which we discussed above, is only one of the
laws that were formulated by Boole. The distributive law clearly holds for the
realms of our experience that are described by classical physics. We would now
like to test this law in the quantum realm.

Electrons, and other quantum entities, possess an intrinsic angular
momentum that we call *spin*. It is tempting to conjure up an image of a little ball
spinning about some axis. Try not to do that if you can help it. The things that we
have learned about the quantum realm suggest that our experience of how
mechanisms function in the world of appearances is not a reliable guide to what
is going on with electrons and photons.

Spin is quantized, which is to say, it can only have certain discrete values. Along any particular axis (x, y, or z), the spin is either *up* or *down* and nowhere in between. Again, do not attach the usual meaning to 'up' and 'down'. They are just labels that denote the only two values that the spin can have. Quantum mechanics tells us that we cannot specify spin simultaneously along more than one axis.

In Fig. 10-1, we show a device first used by O. Stern and W. Gerlach in 1922 to separate a beam of atoms into its two spin components. This is done by introducing the beam into an inhomogeneous magnetic field, which causes the atoms with spin-up and the atoms with spin-down to be deflected in opposite directions. Having shown how spins are sorted, look now at Fig. 10-2, which shows the results of successive spin sortings. First we sort electron spins along the x-axis. Discarding the x spin-down electrons, we introduce the x spin-up electrons into a second x spin sorter. As we might anticipate, this second sorting confirms that we have only x spin-up electrons. No surprises there. Next, we introduce our x spin-up electrons from the first sorting into a sorter that operates along the y-axis. This sorting produces two beams as expected, a y spin-up beam and a y spin-down beam. We dump the y spin-up beam and introduce the remaining beam, which we would classically assume to contain x spin-up and y spin-down, into another x spin sorter. If this were a classical situation, we would expect to see only an x spin-up beam emerge from this second x spin sorting, just as in the first example. However the result of our having acquired knowledge of the spin along the y-axis is to destroy all spin attributes (not merely our knowledge of them) along the x-axis. Hence, we find both x spin-up and x spin-down electrons emerging from this second x spin sorting.

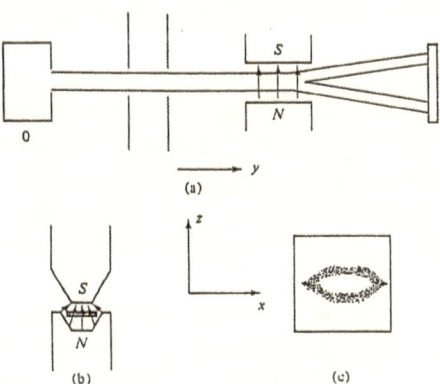

Figure 10-1 A schematic view of the Stern-Gerlach experiment for sorting an electron beam of random spins into two spin-polarized beams along the z-axis, one up and one down.

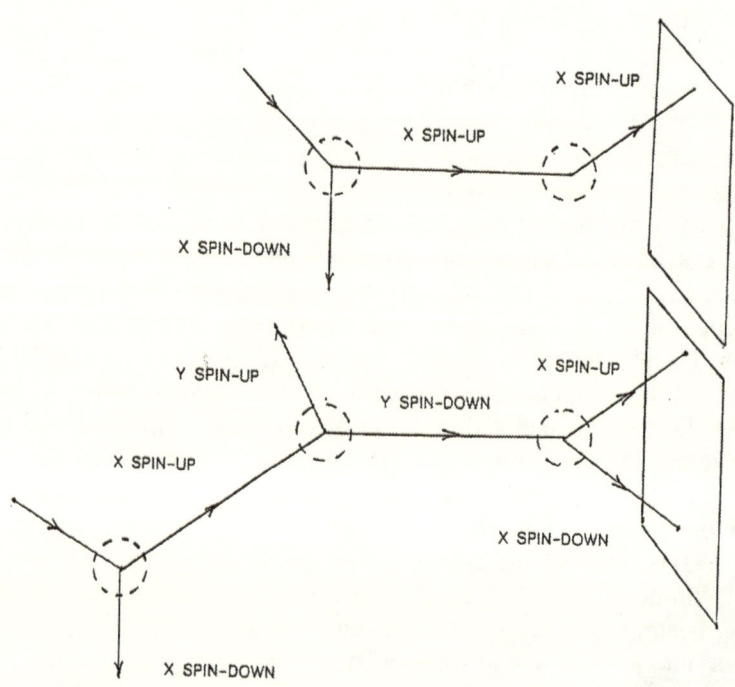

Figure 10-2 The observed properties of spin.

Now let's test our observation according to the distributive law of logic with which we opened this chapter. The statement reads: y spin is down and x spin is either up or down implies that y spin is down and x spin is up or y spin is down and x spin is down. As we have just seen, this does not hold true. The central question to be answered in this chapter is the following. Does the statement not hold true because y spin is down and x spin is (simultaneously) either up or down, but the logic is inapplicable (logical weirdness), or is the initial statement false, thus falsifying the logically valid conclusion (physical weirdness)? We have to know the answer to this question in order to know what we are dealing with.

In 1932, John von Neumann at the Institute for Advanced Study in Princeton and Garrett Birkhoff, a Harvard mathematician, began to study the new field that they called quantum logic. In 1936, they published the first paper on the subject in which they showed that particles combine with the familiar Boolean logic, whereas quantum entities, being represented by wave functions, combine with an altogether different *wave logic*.

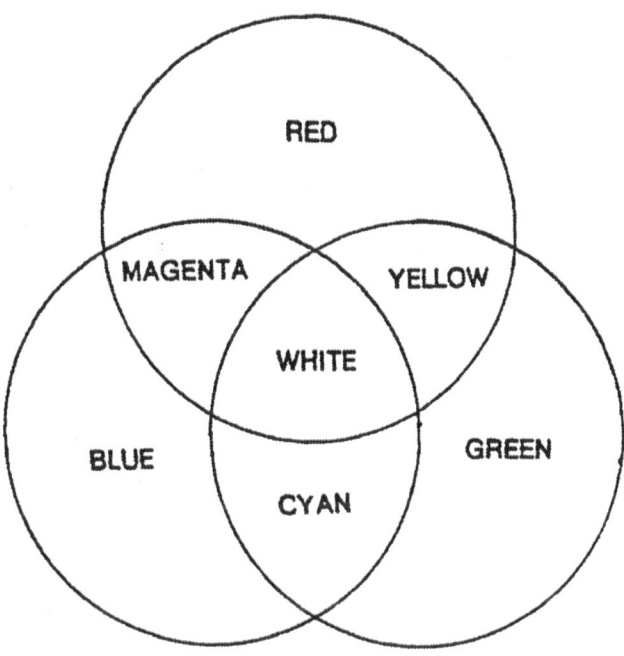

Figure 10-3 Color addition by the mixing of colored lights.

Let us now examine a set of attributes that combine according to Boolean logic. For our example, we will take color. Color can be mixed in either of two ways. There is *additive* color mixing, which occurs when spots of different colored lights are shone on a reflective screen and allowed to overlap. This is seen in Fig. 10-3, where we show the primary additive colors: red, green, and blue. The overlap of red and blue forms magenta. The overlap of blue and green forms cyan (turquoise). The overlap of red and green forms yellow. Where all three primary lights overlap, we get white.

The colors we get when we mix paint obey different rules. There, the process is selective absorption and the *subtractive* primary colors are cyan, magenta, and yellow. We have seen that when you mix red, green, and blue lights, the result is white. On the other hand, if you mix red, green, and blue paint, the result is a muddy brown.

We can sum up all of this information on how color attributes combine by displaying the logic in what mathematicians call a *lattice*. Figure 10-4 shows an example of a Boolean lattice representing color attributes. The lattice is a diagram that displays the vertical ordering of the attributes' inclusiveness, from all inclusive (white) at the top to entirely exclusive (black) at the bottom.

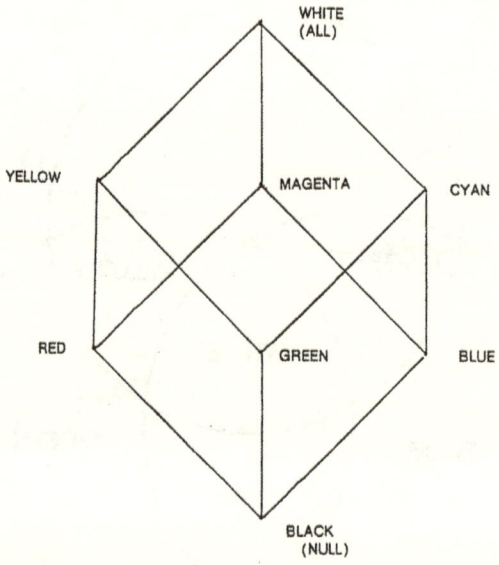

Figure 10-4 Boolean lattice showing the combination of color attributes according to the operators "or" and "and".

The rule for determining the result of applying the "and" operation to any pair of color attributes, say yellow (Y) and magenta (M), is to look for the *highest* common attribute that can be reached by following lines *downward* from Y and M. In this case, the answer is red (R), the most inclusive attribute that yellow and magenta have in common. By using this method, we can apply the "and" operation to any pair of attributes in the lattice.

In order to apply the "or" operation to any pair of color attributes, for example, blue (B) or green (G), we look for the *lowest* common attribute that can be found by following lines *upward*, in this case, cyan (C). The "and" operation is used in stacking color filters and in the additive mixing of light or television images. The "or" operation is used in mixing paints.

Color is a typical example of combining attributes according to Boolean logic. This can easily be shown.

$$Y \text{ and } (R \text{ or } B) \rightarrow Y \text{ and } M \rightarrow R$$

and

$$(Y \text{ and } R) \text{ or } (Y \text{ and } B) \rightarrow R \text{ or } N \rightarrow R$$

98

so the distributive law of Boolean logic applies. Now let's look at an example of a non-Boolean lattice.

In Fig. 10-5, we see the logic lattice corresponding to the combination of *polarization* attributes. In this case, the attributes are horizontal polarization (H), vertical polarization (V), diagonal polarization (D), and cross-diagonal polarization (D*). The rules for the application of the "and" and "or" operators are the same as in the Boolean lattice. "All" at the top of the lattice is the most inclusive and "Null" at the bottom of the lattice is the least inclusive. Quantum lattices, of which the polarization lattice is an example, obey all of the Boolean rules except for the distributive rule. For example

$$\text{H and (D or D*)} \rightarrow \text{H and A} \rightarrow \text{H,}$$

whereas

$$\text{(H and D) or (H and D*)} \rightarrow \text{N or N} \rightarrow \text{N,}$$

so the distributive rule fails in this case. The full lattice (A, H, V, D, D*, N), as shown in Fig. 10-5, is non-Boolean, however the sub-lattices (H, V, N, A) and (D*, D, N, A) *are* Boolean and are quaintly referred to as *Isles of Boole*

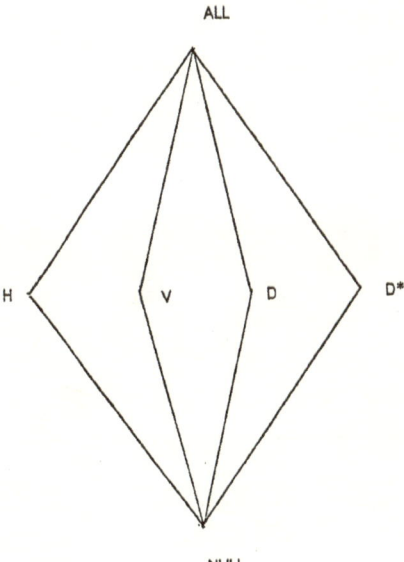

Figure 10-5 Non-Boolean lattice showing the combination of the polarization attributes according to the operators "and" and "or".

In a classical object, all attributes can be measured simultaneously. This is what makes them Boolean. The Boolean sub-lattices are likewise composed of compatible attributes such as H and V that can be measured simultaneously.As a practical example of the relations between quantum attributes, one that displays quantum wackiness in all its glory, we consider the *three-polarizer paradox*.

Figure 10-6(a) shows a stack of two polarizers, one with its axis aligned vertically (V) and the other horizontally (H). The light coming through the V polaroid is vertically polarized and is then unable to pass through the H polaroid. Put a light behind the stack and peer into it from the other side and it appears opaque. Now place a diagonally oriented polarizer (D) *between* the V and the H as in Fig. 10-6(b) and suddenly, the light shines through and the stack becomes transparent. Even if you know no physics at all, your body will get the tingles to let you know that you are seeing something strange. Our experience tells us that if two filters block all the light, adding another one can hardly improve matters. But it does.

The trick only works if the D polarizer is placed between the V and the H. The order DVH and the order VHD will both be opaque. Only the order VDH (or HDV) will pass light. Again, our experience with filters is that the order is immaterial. This is a bit of naked quantum weirdness intruding itself into the world of appearances, and it is startling.

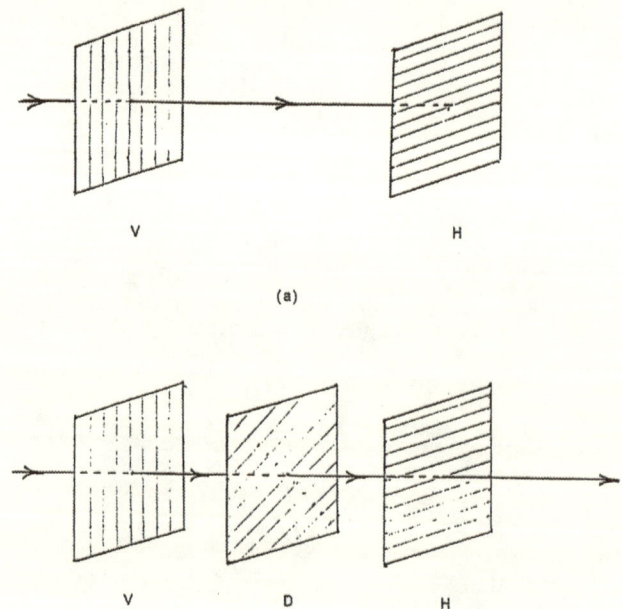

(a)

Figure 10-6 The three-polarizer paradox.

In the classical world of everyday appearances, the order in which we stack filters and the order in which we multiply numbers does not matter. We say that such operations *commute*. In quantum mechanics, not all operations commute. The result of applying the position operator **Q** followed by the momentum operator **P** is not the same as first applying **P** and then **Q**. We say that **P** and **Q** are non-commuting operators. It is the same with the polarization operators. The operators H and V commute; it makes no difference the order in which you stack them. Neither H nor V commutes with D or D*, which do commute with each other.. Hence the result of stacking polarizers in the order DV is not the same as stacking them in the order VD. This is not one of those explanations where you can 'see how the little men are running around down there', but that is usually the case with quantum phenomena.

So now let's get back to the serious question that we raised earlier. Which is the real issue, logical weirdness or physical weirdness? Boolean logic applies in the world of appearances (by and large), whereas non-Boolean logic applies to quantum phenomena. Several authors have stressed the analogy with geometry. Euclidean geometry seems to be a feature of our world of common experience, but, as Einstein showed, it gives way to non-Euclidean geometries in the limit of great distances and/or large concentrations of mass-energy-momentum. Can we use this analogy to claim that quantum attributes really are objective if we apply the correct logic? I think not. There is a distinction between these two examples that renders the analogy false.

In general relativity, we have a correspondence principle. General relativity goes over seamlessly into Newtonian gravitation and special relativity as lengths become moderate and concentrations of mass approach molecular as opposed to nuclear densities. The same is not true with logic. There is no correspondence principle for logic. It's either Boolean or non-Boolean all the way. Clearly the world is overwhelmingly Boolean in our experience. Non-Boolean logic only appears to arise in quantum phenomena and by giving up the notion of determinism and objectivity we can create a consistent picture of what's going on and still retain the universal validity of Boolean logic. Some people would say that this is too high a price and that to give up determinism and objectivity is to invite man's oldest nemesis, chaos, in. I will strive mightily, using psychological, philosophical, and religious arguments to try and convince you that the implications of quantum mechanics are both liberating and uplifting.

Boole may have been knocked off his feet and be down for the count, but, to my way of thinking, he has been saved by the bell. Speaking of Bell, we will introduce him in the next chapter.

CHAPTER ELEVEN

SURELY ONE AND ONE STILL MAKE TWO

in which the phrase "staying in touch"
takes on a whole new meaning.

As you may recall, Albert Einstein, with his thoroughly classical mindset, had resisted the metaphysical implications of quantum mechanics from the very beginning. He and Niels Bohr had an ongoing public dialogue on this subject for many years. They would meet at a physics conference and Einstein would present Bohr with a thought experiment that seemed to show that some facet of quantum mechanics (usually the uncertainty principle) was in error. Bohr would then experience a few hours of panic until he could manage to find the flaw in Einstein's argument, which was always rather subtle as you might expect. It went on like that for years. Gradually, as the successes of quantum mechanics accumulated, Einstein went from claiming that quantum mechanics was wrong to suggesting that it was incomplete. In 1935, he published a paper with Boris Podolsky and Nathan Rosen in which he presented his ultimate argument. This paper (hereinafter referred to as EPR) used the non-commuting operators **P** and **Q** to state their case. David Bohm recast EPR's argument in terms of spins, which makes the discussion somewhat easier to follow. Here is the situation.

An event somehow generates two particles whose total intrinsic angular momentum is known to be zero. The details don't really matter. We could have a pair of low-energy photons emitted by a transition to the ground state in an atom such that one of the photons must be spin-up and the other spin-down. We could consider the high-energy photons resulting from the antimatter annihilation of electrons and positrons, or pairs of protons in the singlet state. Any of these will do. The point is that the two particles that make up the pair are described by a common wave function that is *definite* concerning the total spin of the two-particle system (zero), but is *indefinite* as to the spin of either member of the correlated pair. The two members of the pair travel in opposite directions until they achieve a macroscopic separation, meters or light-years, it doesn't really matter. The fact that the two particles are in a singlet state means that when the spin of one of them is measured to be, say, spin-up, the wave function must *instantaneously* collapse and actualize the spin of the other to spin-down, no matter what the separation between them.

EPR's objection to this was that unless the situation was deterministic, that is, unless one of the particles was spin-up and the other spin-down *all along*, then there was the implication of non-locality (spooky action at a distance), which would appear to violate special relativity. EPR expressed the options by stating that the assumption of the simultaneous validity of the following three statements leads to absurd conclusions: First, quantum mechanics is correct; second, quantum mechanics is complete; third, the results of measurements on microscopic systems are determined by *elements of reality* associated with the measured system and/or the measuring apparatus that remain unaffected by measurements on other distant microscopic systems.

This last statement is equivalent to the assumption of *local hidden variables*, deterministic elements whose influence can only propagate at sub-luminal velocities.

By 1935, everyone, including Einstein, knew that quantum mechanics had an unbroken record of giving correct answers. The first assumption was therefore regarded as true by everyone on both sides of the issue. Einstein thought that the completeness assumption was at fault, whereas Bohr and the Copenhagenists reckoned that the problem lay with the third statement, the assumption of local hidden variables. After all, the great John von Neumann had stated in his book, the quantum bible, that all hidden variable theories, local and non-local alike, were inconsistent with quantum mechanics.

It was generally appreciated that Einstein, Podolsky, and Rosen had raised an important issue, one that had not been satisfactorily settled. No one was a hundred percent sure that quantum mechanics could be applied over macroscopic distances and until some definitive data or clarifying argument was forthcoming, the question would just have to be shelved.

In 1952, David Bohm, ignoring von Neumann's prohibition, wrote a hidden variable theory that seemed to reproduce the predictions of quantum mechanics in their entirety. Bohm's hidden variable theory was non-local, however. Neither Bohm nor anyone else quite knew what to make of that. Special relativity had passed all of the tests to which it had been subjected as had quantum mechanics, but unless one was prepared to believe that an effect could precede its cause, something had to give. The EPR paradox persisted.

In 1964, John Stewart Bell, a very talented physicist who had grown up in Belfast, Northern Ireland, was on sabbatical from CERN, the European Laboratory for high-energy physics in Geneva. In that year, he discovered the error in von Neumann's statement that hidden variable theories were inconsistent with quantum mechanics. Of even greater importance, he formulated a mathematical condition that was capable of distinguishing between the outcome of an EPR experiment that one would expect on the basis of local hidden variables and the outcome predicted by quantum mechanics. The two outcomes were definitely not the same. His argument was very simple and elegant, and I will explain it to you in detail without having to impose on your good nature in order to do so.

The EPR experiment consists of an atomic source capable of emitting correlated singlet-state photon pairs. Using filters, apertures, and lenses, the two photons are sent in opposite directions as shown in Fig. 11-1. In the path of each photon, we place a calcite crystal.

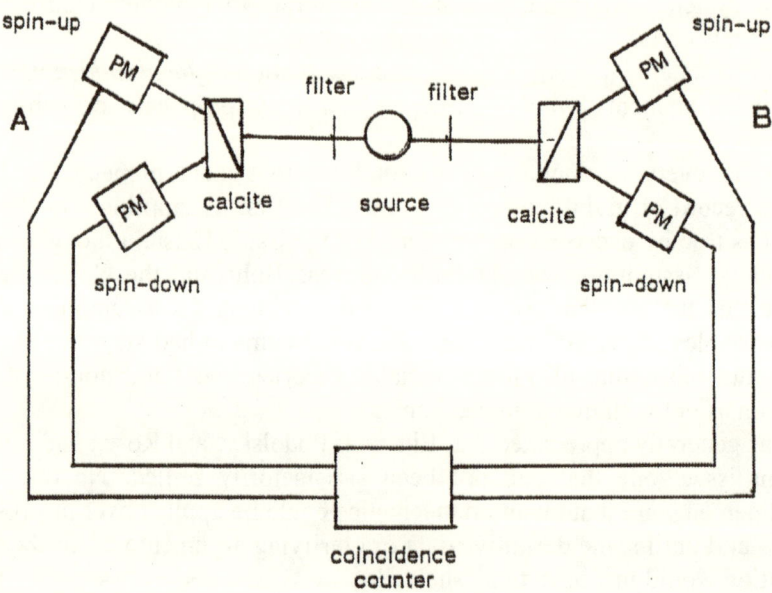

Figure 11-1 Schematic representation of the EPR experiment for measuring the polarizations (spins) of two correlated photons by means of calcite crystals and photomultipliers. The coincidence counter is necessary in order to identify signals from both members of a correlated pair as opposed to signals from single photons.

Calcite (or Iceland spar) is a transparent mineral that has the property of *birefringence*, which is to say that it bends light in two ways, depending on how the light is polarized. An unpolarized beam of light will be broken into two beams, one of whose photons will all be spin-up with respect to the axis of the crystal, and the other, spin-down.

The calcite crystals are arranged so that their optical axes are oriented perpendicular to the direction of motion of the photons. The calcites are capable of being rotated so that any desired angle between the optical axes on either end can be set.

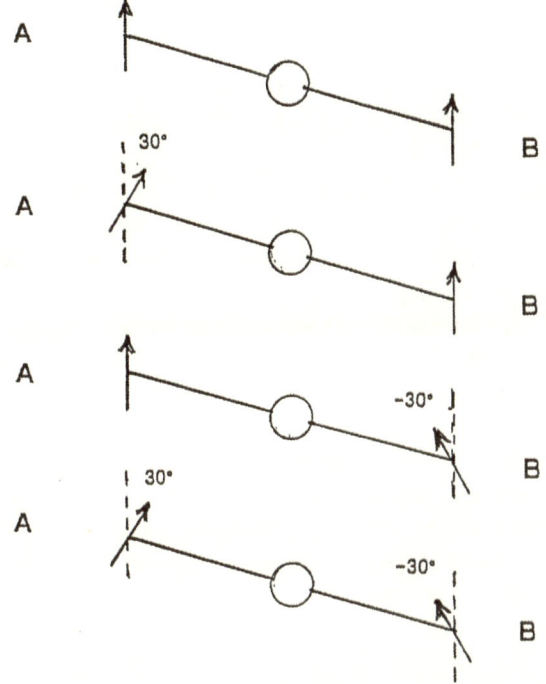

Figure 11-2 Illustrating how the simplest Bell inequality comes about.

Look now at Fig. 11-2. At the top of the figure, we see the calcites at both ends of the experiment set with their axes straight up. With this setting, the correlation is 100 percent. Perfect correlation for singlet-state photons means that every time you measure a spin-up at A, you will measure a spin-down at B, and vice versa. A typical set of measurements for this configuration might look as follows:

A: U D U U D U D D U D U U U D D D
B: D U D D U D U U D U D D D D U U U

Perfect correlation for the s-state is really anti-correlation but let's not confuse things with more terminology than we really need.

Looking at the next step in Fig. 11-2, we set the calcite at the A end of the experiment at an angle of, say, 30 degrees to the vertical, leaving the calcite at the B end at zero degrees. When we run the experiment now, we find that one out of four pairs fail to correlate, that is, they fail to give opposite readings. The data from this run might look like:

A: U D U U D U D U U D U U U D D D
B: D U U D D D U U D U D D D D U U

Next, we move calcite A back to the vertical and set calcite B at minus 30 degrees. Again, we see a failure to correlate of one part in four as we might expect. After all, the two ends of the experiment are identical. Whether we misalign calcite A by 30 degrees or misalign calcite B by 30 degrees should make no difference to the result, and indeed, it doesn't.

Now, let's set calcite A at 30 degrees *and* set calcite B at minus 30 degrees. What should we expect now? Bell's inequality tells us.

Since the failures to correlate at the left-hand end are one part in four for a setting there of 30 degrees and the failures to correlate at the right-hand end are one part in four for a setting there of 30 degrees, the failure to correlate when both calcites are set to 30 degrees for a total angular spread of 60 degrees is, at most, two parts in four, and may be less owing to some of the reversals canceling out.

This is Bell's inequality in a nutshell. Note, however, that quantum mechanics predicts a *three* parts in four failure to correlate for this case, so either the above, very straightforward logic is somehow wrong and one plus one really does make three, or quantum mechanics is wrong.

Let's look closely at the assumptions that go into the Bell inequality. What Bell is saying is that how you chose to set the calcite at A can only affect the outcome at B (if at all) in a time greater than or equal to the time that it takes light to pass from A to B. In other words, the issue here would seem to be *locality* (or the lack of it). Clearly, a real (as opposed to 'thought') experiment was needed to decide the question.

Up to this point, the EPR experiment was still a thought experiment, but by the early 1970s, with the impetus of Bell's inequality condition, it became do-able.

Stuart Freedman and John Clauser at the University of California at Berkeley carried out the experiment in 1972 using low-energy photons emitted from calcium atoms. Their results were in agreement with the predictions of quantum mechanics, that is, the results violated Bell's inequality. In the next four years, there were six more experiments, four of which agreed with quantum mechanics and two that agreed with Bell's inequality. This score of five to two in favor of quantum mechanics is actually a much stronger result than the ratio of five to two would imply. The refinement of the experiments favoring quantum mechanics was generally greater than that of the experiments that failed to violate Bell's inequality. Systematic flaws in the experiments would tend to destroy evidence of a real correlation and would, therefore, favor Bell's inequality over quantum mechanics. Most of the workers in the field felt that the predictions of quantum mechanics had been upheld, but everyone was looking forward to a truly definitive experiment. It was eight years in preparation, and when it was finally run in 1982, it violated Bell's inequality in no uncertain terms. John Bell's *reducto ad absurdum* proof that no local hidden variable theory can reproduce the results of nature and quantum mechanics was complete.

The five successful experiments carried out in the 1970s had shown that results were independent of the distance separating the two measuring devices from the source and whether or not the distance from each measuring device to the source was the same.

There was still the objection that the calcite settings were constant during the experiment, allowing time for communication, by whatever means, between the two ends. This did not seem a likely source of error, but who's to say in this business? It would clearly be a comfort to be able to vary the calcite setting while the photons were in flight so as to remove this last possible source of error and make the experiment definitive.

The French are the absolute world leaders at experimental optics and always have been. In 1982, Alain Aspect, J. Dalibard, and G. Roger of the Institute of Theoretical and Applied Optics of the University of Paris-South in Orsay carried out the experiment depicted schematically in Fig. 11-3. In this experiment, the photons first encounter fast optical switches that periodically (every ten nanoseconds) vary the paths that the photons will take. Ten nanoseconds is the time that it would take the photons to travel three meters. Each switch is a small water-filled chamber in which standing waves are generated ultrasonically. If the standing waves are turned on, the photon will be deflected to an analyzer that is oriented one way. If the standing waves are turned off, the photon will travel straight through to an analyzer oriented another way. Signals from all four channels are fed into a four-way coincidence counter so as to separate the correlated signals from stray singles. The distance between the analyzers is 13 meters, so in the 40 nanoseconds that it would take light to travel between them they are effectively precluded from carrying on a coherent conversation by any means whatsoever. The results of the Aspect experiment violated Bell's inequality quite decisively. It would seem that the issue is no longer in doubt.

I direct your attention back to the discussion of the consequences of super-luminal information propagation at the end of Chapter Four. In essence, what we found was that if a signal is sent at some time, say $t = 0$ at a speed u greater than c, then for some observer moving at speed v with respect to the frame in which the signal is propagated, the signal will be detected at a time t' less than zero.

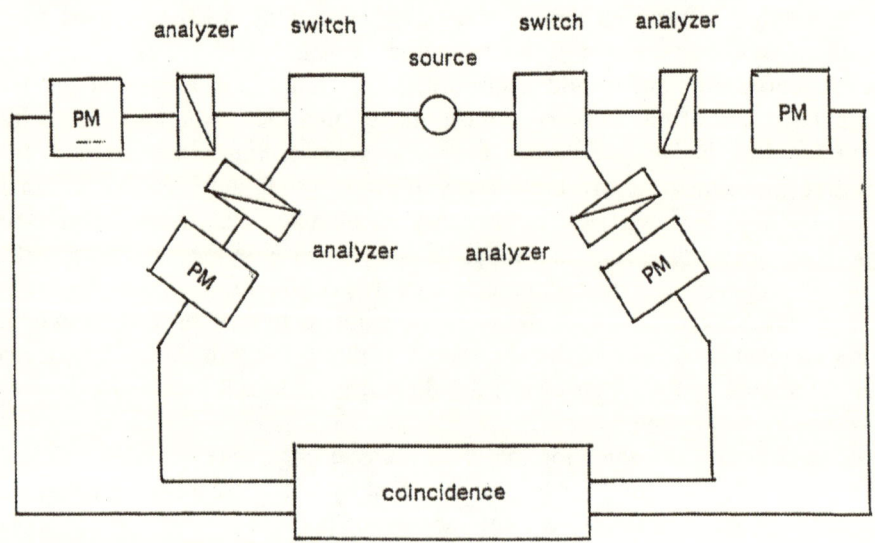

Figure 11-3 The Aspect Experiment.

Specifically, this will hold if $uv > c^2$. This amounts to a reversal of cause and effect, an effect that occurs before its cause. We have already seen this in the delayed choice experiment discussed in Chapter Eight. We will discuss this aspect of non-locality further in the next chapter in a way that will allow us to explore some of its consequences. For now, let's talk a little more about hidden variables.

First of all, let me state again that hidden variables are not a scientific theory. They do not make any predictions that quantum mechanics cannot make on its own. The entire reason for postulating hidden variables is to save determinism, a philosophical position that some people feel must be preserved at all cost.

We have seen that John von Neumann claimed to have proved in his book, *The Mathematical Foundations of Quantum Mechanics*, that all hidden variable theories are inconsistent with quantum mechanics. De Broglie, and later David Bohm, had written hidden variable theories anyway and they seemed to work. In 1964, John Bell had come along and found a flaw in von Neumann's proof, which seemed to legitimize hidden variable theories. In that same year, Bell had written his inequality, which suggested that only non-local hidden variable theories could be consistent with quantum mechanics.

In 1978, F. Selleri and G. Tarozzi published a paper (Il Nuovo Cimento **48**, 120) in which they showed that Bell's inequality is satisfied by the two best-known *non-local* theories, namely Newtonian dynamics and the de Broglie-

Bohm hidden variable theories. This was before the Aspect experiment and Selleri and Tarozzi suggested that maybe this meant that the quantum mechanical theory of correlated spins was wrong. Well, the Aspect experiment has come and gone and quantum mechanics is still intact, so we have to ask ourselves what Selleri's and Tarozzi's result really means. What it would appear to show is that non-local as well as local hidden variable theories are inconsistent with quantum mechanics. The key issue here would appear to be determinism instead of locality.

In one year, John Bell raised hidden variables from the grave in which von Neumann had buried them, and then wrote his famous inequality, which killed them off again. I'm sure he would have appreciated the irony, and I am only sorry that he did not live to experience the joke.

CHAPTER TWELVE

THINGS THAT GO BUMP IN THE NIGHT

in which we cautiously dip a toe into the
murky waters of parapsychology in
search of the quantum connection.

I realize that I run a grave risk of losing credibility with my readers by getting within a country mile of parapsychology. Up to this point, I have stuck to science, which has definite rules by which it conducts its business. Science requires that its theories be submitted to the test of experiment. The trial may be determinant - do this, get that - or it may be statistical. Conduct an honest coin flip a zillion times and get half heads, half tails. What you don't want and cannot abide in science is *experimenter effect*. That is a dirty word. Parapsychology is, of course, wall-to-wall experimenter effect and it is no wonder that the subject turns most physicists off so badly. The evidence for parapsychology wanders all over the map with variations in personality types, mood, motivation, and ego involvement.

There is an area of parapsychological research that might be called *quantum psychokinesis*. This area has direct bearing on the nature of the reality that underlies our world of appearances. Just as the coin-flip probability averages out to 50:50 if the data base is sufficiently large, so in quantum psychokinesis we can average out the experimenter effect with a large enough number of runs and produce data that seems capable of meeting science's criteria for replicability.

There is another reason that I dare to bring this up. I have been personally involved in it. You may remember that Nicholas Copernicus came up with his heliocentric view of the cosmos, but waited until he was on his deathbed and safely beyond the reach of the Inquisition to publish. Not to compare myself to Copernicus, but I am taking somewhat the same approach. I did the work that I am going to describe to you in 1982-1983. At that time I was still trying to advance an academic career and I was well aware that a publication on the subject of parapsychology would not exactly enhance my resume. I did the work because I was curious. I satisfied that curiosity and sat on the results until now, when career considerations no longer motivate me.

In 1981, I became aware that Helmut Schmidt of the Mind Science Foundation in San Antonio, Texas was carrying out experiments in which he claimed to show statistical evidence for the manipulation of quantum mechanical probabilities by human intentionality. Let me clarify that a bit.

We know that half of a given sample of radioactive material will decay in a certain time, which is known as the *half-life* of that particular isotope. We do not know, however, the times at which individual atoms will decay. That, according to quantum mechanics, is an altogether random event. Schmidt claimed that by employing what I might call a suitable *psychological lever*, a human subject could bias those decay times. The details are shown in Fig. 12-1.

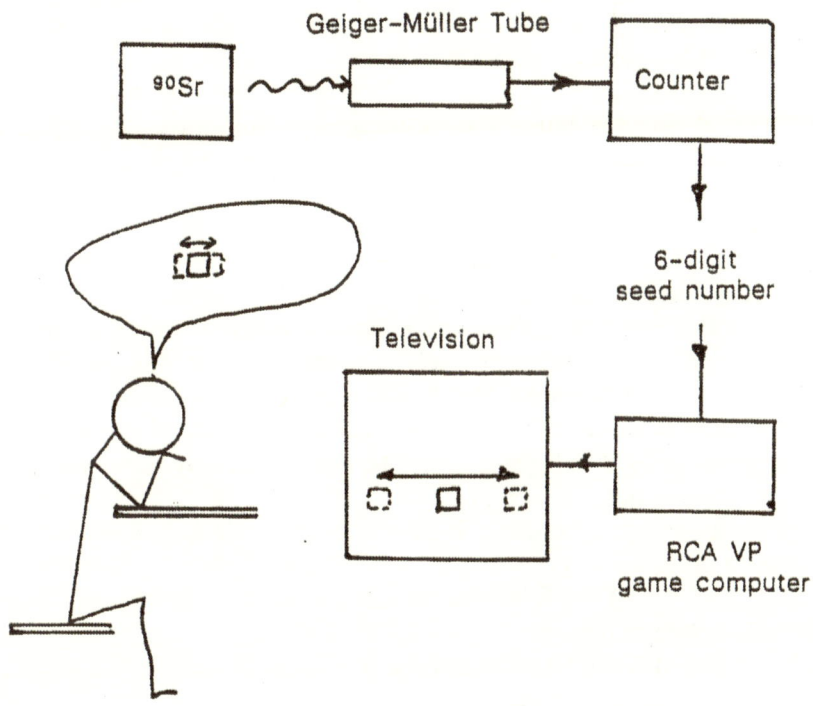

Figure 12-1 The quantum mechanical psychokinesis experiment.

A decay event in a radioactive element, in this case strontium-ninety, triggers a Geiger-Müller tube at intervals that quantum mechanics holds to be entirely random. The output of the Geiger-Müller tube triggers a running counter and, in that way, generates a 6-digit number whose randomness is of the same order as the randomness of the decay. This 6-digit 'seed number' is fed into a computer. The computer is programmed with a long closed loop of quasi-random (that is to say, algorithm-generated) 6-digit numbers. The seed number from the nuclear decay determines the starting position in the loop for a one-minute run. The output of the computer uses the randomly selected segment of the loop to control the amplitude variations of a square linear pendulum oscillating back and forth on a television screen. In this way, the dynamic behavior of the linear pendulum is linked directly and unambiguously back to the time of the nuclear decay event. This linear pendulum and its dynamic behavior constitute the psychological lever.

In front of the television screen is seated a human subject who concentrates on willing the varying amplitude of the pendulum oscillation to stay as small as possible. The subject tries, in effect, to clamp the image of that square with his or her mind.

As odd as it may seem, Schmidt had statistical evidence that the human subject's effort appeared to make a difference.

I was interested in the conceptual underpinnings of quantum mechanics then as I am now. I found Schmidt's work (if true) mildly intriguing. While I was dithering around, trying to decide whether or not to pursue this, Schmidt added another wrinkle. The experiment, up to that point, had operated in 'real time' with the decay-generated numbers fed directly to the computer. For the purpose of dealing with me at a distance, he proposed to record the 6-digit seed numbers for later use, thereby enabling some statistical analysis to be carried out on the seed numbers. There was, of course, already a delayed-choice feature in the experiment, and it did not seem likely that the use of pre-recorded numbers would interfere with whatever effect might be present. I was hooked. I got on the phone to Helmut Schmidt in San Antonio and before you could say Joseph Banks Rhine, a primitive RCA VP game computer was on its way to me along with a printout of decay-generated seed numbers.

If this was going to be done, it should be done right. I thought about the experimental protocol very carefully. The first thing I did was to enlist a Critic Committee made up of the full professors in the physics department. They were kind enough to humor me.

When a list of 6-digit seed numbers arrived from Helmut Schmidt in San Antonio, it would be turned over to the Critic Committee. They would randomly select about a third of the numbers to be tested for randomness using standard mathematical tests. For example, the range 000000 to 999999 would be divided into (say) 20 equal intervals and the count of numbers falling into the various intervals would be compared. Another test consisted in looking at the 6N digits in N numbers and comparing the number of odd and even digits. The numbers screened for randomness would then be set aside.

The remainder of the seed numbers would be separated into two batches, one to be used in the experiment and the other as a control. It was established using the algorithm furnished by Schmidt that the mean 3-digit score of a run was 500 just as he claimed with a standard deviation of 123. The success of a run would be judged by whether the score printed on the screen at the end of a run was less than 500 (a hit) or greater than 500 (a miss). The assigned seed numbers were typed into the RCA VP game computer by the subject, who then composed himself for the effort and started the run. The subjects were myself, three of my students, and my son, then aged twelve.

The final third of the seed numbers were used as control by running them through the computer to generate scores without turning on the television display.

The concentration and effort required of a well-motivated subject were considerable. Two or three runs was about as much as any one person could manage in a day. In two years time, we carried out a total of 2092 runs, of which 1113 were hits and 979 were misses. That doesn't sound very impressive, but let me put it in perspective. The hit-to-miss ratio was 1.14. The number of standard deviations from the mean was 2.93. The ratio of the probability of 50 percent hits to the probability of the score achieved was 73.19 to one.

By contrast, the control group of 2092 runs with 1042 hits and 1050 misses showed a hit-to-miss ratio of 0.992; the number of standard deviations from the mean was 0.175, and the probability ratio as defined above was 1.01 to one.

This data convinced me that quantum psychokinesis was a real effect. If you add my scores in with those accumulated by Helmut Schmidt and his subjects over the years, the conclusion, which I knew first-hand to be true, was that the human mind, aided by an appropriate psychological lever, is capable of biasing random, acausal, quantum processes.

Arguments that we have looked at suggest that we all, collectively and unconsciously, manifest the world of appearances by actualizing quantum potentialities. It was worth the effort to find out if individuals can do this consciously. The answer is yes, we can. Does this imply that human beings can mentally manipulate classical phenomena, bend spoons, that sort of thing? Not at all. It would seem, however, that human beings can influence such things as decay times and whether a spin actualizes up or down. That is enough to erase any lingering doubts in my mind about man's role in the scheme of things. The experiment was a lot of hard work, but the knowledge I gained is a matter of experience, not hearsay or theory and that, of course, is the best sort of knowledge.

CHAPTER THIRTEEN

REALLY, JUST FOR ME?

in which we appreciate just how well
suited for our comfort this Universe is.

We need to remind ourselves at this point that our main aim is not merely to entertain ourselves with the wonders of physics, but to try and gain some overall notion of the nature of existence and man's role in the Universe. It has become clear from our discussion of quantum mechanics that we are not simply passive observers, but active participators, endowed with the power to actualize quantum probabilities and promote them to full reality status in the world of appearances. In this chapter, we will examine further evidence to help us clarify just what role we play.

In the West, the place of man in the scheme of things was first elucidated by Aristotle. He placed the Earth at the center of the Universe, with Hell below and the Heavens above, revolving on their crystalline spheres. Thus man stood at the center, suspended between Heaven and Hell, and was regarded as the chief concern of the gods, who, in the Greek pantheon, shared many of man's foibles and neuroses. The medieval Christian Church found that this cosmology suited its purposes and hence, they dogmatized it.

Nicholas Copernicus overthrew all that and set us on the path to mediocrity when he argued that the Earth was just one of several planets orbiting about the Sun, which now assumed the central position in the Cosmos. The modern scientific view is faithful to the essence of Copernicus scheme in its greatly expanded scope, and sees the Earth as but one of nine planets orbiting around a quite ordinary star. This star is, however, not the center of the Universe, but one of a hundred billion in our galaxy, the Milky Way, our galaxy being but one of a hundred billion in the Universe.

The current paradigm, which governs our behavior and defines *common sense*, regards man as an intelligent ape, sadly aware of his own mortality, scrambling to keep himself entertained and distracted from what appears to be the grim facts of life. Most biologists and many physicists regard our minds as mere epiphenomena of our brains, which are themselves no more than supercomputers made of meat. On the basis of this view, man is a machine, a mechanism with no free will of his own. The Church and society expect us to be responsible for our behavior and guilt or punish us when we stumble. In other words, the materialistic paradigm takes away all of the benefits of free will, while our social structure necessarily burdens us with its responsibility. Is it any wonder that we are all a bit schizy.

Secular science has given us remarkable material wealth and a high physical standard of living. It has cured many of our diseases (and given us some new ones). It has not made us any happier or less neurotic, however. By no means all, but certainly the most visible religious activity in our culture operates almost exclusively on the emotional level. Fundamentalist intolerance is on the rise all over the world and is responsible for the spilling of a great deal of blood in the name of God, Allah, or Whoever. It has been said that the last thing a bishop wants to find is a saint in his parish. Isn't that sad?

To sum up this depressing discussion, we note that our paradigm is no longer serving us well. The material paradigm was necessary in order to make the transition from magic to science, but now that this development has led to quantum mechanics, it is time to cut our spiritual loses and move on. The knowledge that we have gained in physics over the last hundred years has killed the basis for the nineteenth century materialistic paradigm that we slavishly follow. We are keeping company with a decaying corpse, a zombie animated only by the force of our habit and the fear of change. The odor of corruption is getting pretty intense. Let's look at one more piece of evidence as to man's true nature and rightful role and then see if we can make some positive decisions.

In 1961, Robert Dicke, a Princeton physicist, wrote a remarkable paper. I remember reading it and thinking to myself that it was probably important. In order to discuss Dicke's paper, I need to back up a little.

As you may recall from Chapter One, the Copernican revolution had the effect of removing man from his central position in the Cosmos. From the Copernican Principle, there developed the Cosmological Principle, which states that the Universe is homogeneous and isotropic, that is, it is smooth overall and looks the same in all directions from any vantage point. This was extended by Hermann Bondi, Thomas Gold, and Fred Hoyle into the Perfect Cosmological Principle, the idea that the Universe has always and will always look just as it does now. This steady-state theory of the Universe, wherein the observed expansion was driven by continuous creation of matter, was undone in 1965 when the radiative remnant of the Big Bang was discovered. We discussed this in Chapter Five, so this is by way of a reminder.

Let's return now to Robert Dicke's paper of 1961. Assuming the truth of the Big Bang model, which was shortly to be confirmed, Dicke addressed the question of why we observe the Universe to be on the order of ten billion years old, the age deduced from the expansion rate. At first thought, you might say that this is just a coincidence. Dicke pointed out that the Universe had to have been old enough to have produced elements as heavy as carbon, the material from which people are constructed.

Carbon was not present in the early Universe. It had to be synthesized in large stars and then distributed throughout space by supernova explosions. Large stars evolve more rapidly than smaller ones like our Sun, so we can say that the time required to cook up the necessary elements and distribute them must have been about a billion years. Add in a few billion more years for planets to form and cool, life to develop, and intelligence (such as it is) to reach the point where such questions can be asked. This sets a lower limit to the age of a Universe that can contain people like Robert Dicke. On the other hand, if the Universe was much older than it is, then most of the nuclear fuel that powers the stars would have been consumed, and the stars would have collapsed into white dwarfs, neutron stars, and black holes. Life as we know it would no longer be possible. Thus the observed age of the Universe is the age that permits the existence of physicists and other sentient beings.

This argument is an example of what has come to be called the *Weak Anthropic Principle*. While the Copernican Principle and its outgrowth, the Cosmological Principle, requires that all directions, locations, and physical manifestations be unprivileged, the Weak Anthropic Principle says no, the observed values of physical constants are restricted to those values that permit the Universe to evolve into an environment that allows the existence of carbon-based life of the human persuasion.

Physics principles are taken much more seriously if they are capable of making testable predictions. The Weak Anthropic Principle, as set forth by Dicke in 1961 could possibly have been used to argue against the steady state theory. It's a shame that Dicke did not attempt to do this. Such an argument would have put the Weak Anthropic Principle on the winning team when, three years later, the primeval background radiation was discovered.

As you might have guessed, there is also a *Strong Anthropic Principle*. As Brandon Carter has stated it, "The Universe must be such as to admit the creation of observers within it at some stage." In other words, not only must the Universe be as it is in order that life exist, we can say that if life did not exist, neither would the Universe.

Here we are endowing the Universe with a purpose. I had better make clear what sort of stretch this requires. Science has always limited itself to asking *how*, and not *why*.

The question *why* properly belongs to *teleology*, the study of the *purpose* of things. Teleology might be said to be a branch of philosophy, but I really think that *why* is a religious question.

The Strong Anthropic Principle is inspired by studying the incredible extent to which the Universe is suited as an environment for life as we know it, and refusing to accept this as a coincidence. The British astrophysicist, Sir Fred Hoyle was sufficiently impressed by the very comfortable (for us) values of the fundamental constants, and by the convenient positioning of the nuclear resonance levels in carbon and oxygen to draw a teleological conclusion: "I do not believe that any scientist who examined the evidence would fail to draw the inference that the laws of nuclear physics have been deliberately designed with regard to the consequences they produce inside the stars. If this is so, then my apparently random quirks have become part of a deep-laid scheme. If not, then we are back again at a monstrous sequence of accidents."

This opinion cannot be proved or disproved. As we have noted, it is essentially a religious opinion rather than a scientific one. It should not be disparaged on this account, however. If we are ever to make progress with what one might call the *Big Questions*, then we are going to have to take an integrated approach and lose our fear of crossing the imaginary boundaries between disciplines.

If we take what we have learned about quantum mechanics into account, then we are led to a version of the Strong Anthropic Principle that John Wheeler has christened the *Participatory Anthropic Principle*. This principle states that: Observers are necessary to bring the Universe into being. The fact that Man has appeared on Earth only in the last microsecond on the cosmic time scale does not, in any way, violate this statement. Recall the cosmic version of Wheeler's delayed choice experiment where our intentionality reached back several billion years to determine whether the photon went around both sides of the galaxy, or only one side. If observation, or, more likely consciousness, creates the Universe, it can certainly create it retroactively right back to the Big Bang.

Let us suppose, if you will, that the Strong Anthropic Principle in either (or both) the teleological form expressed by Hoyle and/or the participatory form laid down by Wheeler is true. For whatever reason, intelligent life *must* come into existence at some appropriate stage of the Universe's history. If this life were to die out at our stage of development, then it would be hard to see why it must have come into existence in the first place. This suggests a corollary to the Strong Anthropic Principle, which has been called the *Final Anthropic Principle*. It states that: Intelligent life must come into existence in the Universe, and once having come into existence, it can never die out. This being the case, intelligent life must eventually become omniscient and omnipotent, completely molding the Universe into an optimal teaching situation in which to realize the true nature of existence. This condition is a state that the Jesuit paleontologist and philosopher, Pierre Teilhard de Chardin called the Omega Point. Hindu or Buddhist mystics might refer to it as the point at which all sentient beings attain Nirvāna. Despite the religious overtones, the Final Anthropic Principle is really a physics statement with no inherent moral or ethical content. The validity of the Final Anthropic Principle is, however, the necessary physical precondition for moral values to arise and exist endlessly in the Universe. It is a statement of religious principle of the least restrictive sort arising out of physics. What could be more marvelous?

This is such an important point that it deserves to be repeated. We have seen that our physical bodies, including our brains, are quantum objects, which are mortal owing to the demands of Darwinian Evolution. That part of us that is responsible for collapsing wave functions is not a quantum object, which is to say that it is not subject to impermanence, which only applies to physical bodies. Thus, on the level of the collective unconscious (and above), we survive the death process, and have all the time it takes to acquire omniscience and omnipotence, not to mention compassion.

Not surprisingly, the Anthropic Principle in all of its forms is widely criticized. The science writer Martin Gardner, a strong proponent of the current paradigm, has said that the Final Anthropic Principle (FAP) is a lot of CRAP (Completely Ridiculous Anthropic Principle), which is hardly a specific objection, more like an emotional reaction.

More to the point, Heinz Pagels has pointed out that the Anthropic principle takes a very parochial view in suggesting that we are the only intelligent life in this enormous Universe or that any other intelligent life must be close to the human pattern. He also claims that the Anthropic Principle does not make testable predictions, and is therefore not a legitimate physics principle. These are serious objections and I will try to answer them.

Let us assume that intelligent life possessing consciousness is required to bring the Universe into existence. This would seem to be implied by quantum mechanics. It is hardly surprising that such a Universe would embody values of the elementary constants chosen so as to make the environment as comfortable as possible for the creating intelligence. We can only speculate as to the existence of other intelligent races, but if we go along with Pagels and assume that there must be a great number of them (which I am inclined to do), then it seems reasonable to suppose that they must have brought Universes that are specific to their needs into existence, Universes that do not communicate with our own. This sounds like Everett's Many Worlds hypothesis all over again, but it is not. Everett's purpose was to preserve some sort of material reality so that physicists could hang on to the old paradigm. Better the Devil you know - that sort of thing. The current proposal simply allows for the likely diversity of sentient beings with differing needs as to how to work out their destinies.

Pagels' other objection had to do with the Anthropic Principle's presumed lack of predictive ability. Even if his objection were true, that would not mean that the Anthropic Principle was not useful. An example of a non-testable principle that is useful to physics is the Principle of Aesthetics. Mathematically beautiful theories are more apt to be true than ones that are not. Paul Dirac once mentioned in a lecture I attended that the mystery of the advance of planetary perihelions could have been solved by assuming that gravity, instead of going like one over r squared, instead went like one over r to the power $1.9976843...$, or something of the sort. Dirac pointed out that nobody would have believed that. It was too ugly. So it was left to Einstein's beautiful theory of general relativity to solve the problem.

Now, let's take the opposite view and argue that the Anthropic Principle does make testable predictions. First of all, it would predict that we will not encounter other intelligent races in the Universe (our Universe) unless they are physically at least somewhat similar to us. Secondly, we note that modern astronomy has shown that our Universe seems to lie on the very border between *open* and *closed*. Is this a coincidence? I doubt it. The observed geometry of the Universe would seem to guarantee a long life as a benevolent environment for intelligent life. The expansion would appear to be rapid enough so as not to collapse too quickly back into a Big Crunch, and not so rapid as to thin out the material for star formation too quickly. In short, we appear to have a Universe that gives us the maximum amount of time to reach Teilhard de Chardin's Omega Point, the best of all possible worlds.

Protagoras said, 2500 years ago that: "Man is the measure of all things." Man may, in fact, be a bit more than that.

CHAPTER FOURTEEN

METAPHYSICS IS NOT A FOUR-LETTER WORD

in which we come to some conclusions
concerning the nature of existence.

Although our story has been told mostly from the point of view of physics, elements of philosophy, psychology, and religion have all crept in. These are the big four, and they overlap to a much greater extent than academics in these various disciplines would ever admit. Classical physics and quantum physics obviously have philosophical implications, differing markedly in the two cases. The fall of objectivity and the inclusion of perception or consciousness into the reality equation invokes questions having to do with psychology. We saw in the last chapter that the Final Anthropic Principle, which arises in physics, makes a statement that is essentially religious in nature. In this chapter, we shall examine the philosophical connection. In the following two chapters, we will look at the psychological and religious aspects.

In writing this book, I naturally read quite a few of the books that addressed the question of the conceptual foundations of quantum mechanics. All of them, without exception, seemed to be desperately concerned to preserve some sort of material reality. The failure of quantum mechanics somewhere along the way would have solved that problem for the naive realists, but that has not happened and I would not advise anyone to hold their breath waiting for the day. As regards the EPR experiment, material reality might be saved if the *Principle of Local Causes* were found to fail. There are two options there. The first is that locality fails. This is prevented for material bodies by Einstein's relativity. We have seen from Selleri and Tarozzi that non-local hidden variables are inconsistent with quantum mechanics as well. This leaves the possibility that something that Henry Stapp calls *contrafactual definiteness* fails. Contrafactual definiteness is the supposition that if things were other than they are, (which, of course, they are not) a definite result would still be realized. This CFD business gets batted around a lot, however I am not at all sure that it has any meaning, things not being other than they are. Giving CFD the benefit of the doubt, the failure of contrafactualness would imply that no result other than the one we see is possible. This is *bootstrap*, an idea of Geoffrey Chew's, popularized by Fritjof Capra, in which the world is seen to be so highly interconnected that only one pattern of events is possible and no alternate choices are available. On the basis of bootstrap, the world is basically superdeterministic. This attempt to exchange analysis (the bedrock of science) for synthesis does not appear to have born fruit.

The other option is to suppose that definiteness fails, which brings us back to Everett's Many Worlds hypothesis, which does not answer all of the questions either and is unattractive as well. There may be a few more candidate 'material realities', but I cannot, in truth, work up any enthusiasm for them.

The real issue is *metaphysics*, the basic nature of existence. The choices here are rather straightforward. There is *monistic materialism*, the philosophy of those who think that all things are 'things', and that spirit, mind, and consciousness are mere epiphenomena of the material brain. Materialism must be deterministic, objective, and local. We have seen that the tested predictions of quantum mechanics falsify all three of these characteristics, so we can say with certainty that the world is not monistically materialistic.

The next choice is Cartesian Dualism, which postulates two categories of existence, mind and matter, neither being reducible to the other. This is certainly a more palatable choice, but it too does not stand up to close scrutiny. Mind and matter obviously interact, and hence must be regarded as exchanging energy. One of the most basic principles of physics tells us that energy is conserved in the material world, which appears to deny the possibility of mind-body interaction. More basically, the material portion of existence is out of step with quantum mechanics, so, with apologies to René Descartes, Cartesian Dualism must be rejected as well.

The only choice left is *monistic idealism*, the philosophy that holds that all physical manifestations are products of the mind, or (perhaps more accurately) of consciousness. This idea plays pretty well in the East, but often seems to engender a panic reaction in Westerners, who appear to feel as though idealism is equivalent to having the rug pulled out from under them.

Think about it for a minute. We know that matter is made of molecules, which are clumps of atoms. The atoms are composed of electrons surrounding a nucleus of protons and neutrons. The protons and neutrons are, in turn, made of quarks. Now, some leading theorists believe that the quarks and electrons are made of one-dimensional *strings*, about a billionth of a billionth of a billionth of a billionth of a meter long, existing in a ten (or maybe eleven) dimensional space. Thus, the only question we are raising is whether or not these strings are material or ideal. How could the answer to that question possibly threaten anyone? Physicists have this burning need to know that the object of their investigations has objective existence out there, even as strings. In this regard, I think that they are indulging themselves. All that science needs in order to function is the ability to put its theories to the test of experiment. That is not going to go away if we find, as it seems we have, that the nature of existence is mental rather than physical.

It is obvious that monistic idealism, lacking as it does determinism, objectivity, and locality, accommodates quantum mechanics most handsomely. Still, it is a tough sell. The common Western reaction to idealism is to dismiss it as totally absurd. The famous Dr. Samuel Johnson (1709-1784) refuted idealism by kicking a large rock with his toe. I can recall seeing a cartoon in which two men are walking down a hill and one is telling the other that the latest theory of physics is that nothing exists except as we perceive it. Down the hill behind them is rolling a large (and apparently highly objective) boulder.

There is a well-known bit of verse that discusses the idealist position with respect to a certain tree in the Oxford quadrangle:

> There was a young man who said: God
> Must think it exceedingly odd
> That the sycamore tree just ceases to be
> When there's no one about in the quad.

To which God replies:

> Young man, your astonishment's odd.
> I am always about in the quad
> And that's why the tree continues to be
> As observed by
> > Yours Faithfully,
> > God.

This verse expresses a concept in the philosophy of the Irish Bishop George Berkeley (1685-1753), who developed the first fully elaborated formulation of idealism in the West. Note that Bishop Berkeley accounted for the apparent permanence of the conscious perception of the tree by asserting that the tree, while ideal, persists in the mind of God. From a secular point of view, we have seen that the wave function of anything as large as a tree spreads imperceptibly from one viewing to the next. In either case, it is still around when there is no one in the quad.

Western philosophers have tried their dead level best to refute idealism, but without success. If you stop and think about it, the independent existence of a tree (or anything else) that is not being perceived by any conscious being is quite impossible to prove. For this reason, if for no other, *all honest scientists should be idealists*. We have seen, however, that classical physics was based solidly on the comforting notion of objective material realism, and old habits die hard. And then, of course, there is the disconcerting notion that no scientist should have to devote his time and effort to studying an illusion.

A large part of the problem in helping Western physicists to surmount the purely conceptual hurtle of idealism is that very few of them (very few of us, I should say) have any training in philosophy or exposure to Eastern points of view. The only recent book that I have read in which an idealist metaphysics is postulated on the basis of quantum mechanics was written by someone who was born and raised in India. Even he seems to hedge what he calls monistic idealism when he says: "...the philosophy (of monistic idealism) does not say that matter is unreal but that the reality of matter is secondary to that of consciousness, which itself is the ground of all being - including matter."

This sounds just a bit dualistic to me, but the problem may well reside in varying interpretations of the terms *real* and *reality*, which almost everyone understands to imply *physically* real and *material* reality. For that reason, I think that I will try to avoid them.

My own interpretation of monistic idealism is very straightforward. I believe that there is no objective material existence. Period. Everything is a mental construct. It is our shared consciousness that provides the means by which we all dream more or less the same dream and reach what seems like a consensus view of the world of appearances.

The Viennese philosopher Philipp Frank saw the implications of quantum mechanics clearly when he wrote in 1957:

"The solutions of Schrödinger's wave equations (ψ-functions) can be interpreted as probabilities; probabilities are however mental phenomena; hence the ψ-function is interpreted as a mental phenomenon that happens in a human mind; the hydrogen atom is described by ψ-functions; hence the hydrogen atom is a mental phenomenon and is a product of spiritual powers. The case against materialism is proved."

The very best explanation of what I have been trying to say was written by a physicist, Sir Arthur Stanley Eddington in 1927, when quantum mechanics was still quite new and was understood by almost no one:

"The recent tendencies of science do, I believe, take us to an eminence from which we can look down into the deep waters of philosophy; and if I rashly plunge into them, it is not because I have confidence in my powers of swimming, but to try to show that the water is really deep.

To put the conclusion crudely - the stuff of the world is mind-stuff. As is often the way with crude statements, I shall have to explain that by 'mind' I do not here exactly mean mind and by 'stuff' I do not at all mean stuff. Still this is about as near as we can get to the idea in a simple phrase. The mind-stuff of the world is, of course, something more general than our individual conscious minds; but we think of its nature as not altogether foreign to the feelings in our consciousness. The realistic matter and fields of force of former physical theory are altogether irrelevant - except in so far as the mind-stuff has itself spun these imaginings."

I can think of nothing to add to that.

CHAPTER FIFTEEN

IT'S ALL IN YOUR MIND

in which psychology is invited to join in the study of consciousness and its powers to manipulate wave functions.

Physicists occasionally talk to mathematicians, chemists, and philosophers, but none of them talk to psychologists. The split between psychology and physics is old and deep, going back to the time of Galileo Galilei. In his writings, Galileo distinguished properties such as geometric shape, motion, number, hardness, and the like as primary objective properties. Shades of color, temperature, musical pitch, taste and such as that were described as secondary subjective properties, mere reactions of the perceiver. Later on, the philosophers René Descartes and John Locke voiced similar opinions. Impressions produced by the mind were clearly secondary and less significant than properties belonging to the objective material universe. Physics claimed the primary quantities for itself and left the secondary rubbish for the (as they saw it) less reputable pseudo-science of psychology.

As we know, physics has prospered brilliantly from the time of Galileo to the present day. Psychology, on the other hand, has stumbled forward somewhat awkwardly here in the West, with no grand theories or impressive fundamental laws to its name. In many ways, the situation has been quite the opposite in the East. There, psychology has flourished for thousands of years while interest in physics has been minimal.

This is not to say that Western psychologists are stupid. They were, after all, left with material to analyze that did not lend itself easily to the ideal interplay of theory and experiment that physics had established as the *scientific method*. The psychologists almost drove themselves mad trying to apply the methods of physics to the ephemeral and evanescent material that was the object of their investigations.

The worm has turned. We now know that nothing exists independently of consciousness, and so the psychologists have just as much right to play in the big kids' sandbox as the physicists and the philosophers. All of physics is certainly a product of the mind (as is psychology and philosophy), so the academic barrier that has excluded psychology is not only unfair, it is downright counterproductive.

If we are going to understand our world more deeply than we do now, we are going to have to come to serious grips with consciousness. This is a very difficult area to investigate. In many ways, it is like a knife attempting to cut itself. If we are to have any hope of making useful progress, we will have to bear closely in mind the admonition of William of Ockham, the fourteenth-century English philosopher, that in formulating any theory, assumptions must be ruthlessly minimized. This rule is called *Ockham's Razor*.

Physics is certainly not in the position to tackle this problem without the active participation of the best that psychology, philosophy, and religion have to offer. With that in mind, let us begin at the beginning and see what we can infer about consciousness on the basis of introspection and the things that quantum mechanics has taught us.

Up to this point in the book, we have been moving right along, analyzing data, reaching conclusions, and feeling pretty good about the process. In so doing, we have gradually wandered out of the familiar (if problematic) material world of appearances and into the world of mental creations where we are less sure of the rules. I am anxious not to let our exploration, which has maintained a certain standard up to this point, degenerate into a mere mouthing of mystical platitudes, or otherwise run off the track. It is clear, however, that we are going to have to shift our approach, studying our subject in sideways glances so to speak. To continue with the direct approach would both dazzle and mislead us.

Obviously, we have a subjective consciousness of the external world and an internal consciousness of thoughts, feelings, and concepts. We saw in Chapter One that in prelinguistic times the internal consciousness was largely limited to instinctual feelings. Only gradually did the subject-object split occur and full-blown ego awareness emerge - 'me' in here and 'other' out there. These are the aspects of consciousness upon which we can all pretty well agree.

We are also aware, but more dimly, that there is a sub-consciousness or depth-consciousness from which thoughts, dreams, and imaginings arise. Those people who have had telepathic experiences and stop to consider the implications will realize that this subconscious level is a shared thing and is transcendent - beyond space and time. I'm not really talking about anything very spooky here. How many people have been thinking of something or someone, and had someone with whom they are close voice the thought? It is a very endearing experience and not at all unusual. In prelingual times, almost all communication may well have been via this channel, and this may be the case for the higher animals today.

As a further exploration along these lines, let's reconsider Schrödinger's cat as inspected by Eugene Wigner and his friend. We have seen that both Wigner and his friend are equally empowered to collapse wave functions. Suppose now, that Wigner and his friend inspect Schrödinger's cat simultaneously. Do they get into an argument over whether the cat should be actualized dead or alive while the poor cat continues in his half-live, half-dead state, waiting for them to reach a decision? Of course not. It is their perception and awareness that *triggers* the collapse of the cat's wave function, but the decision (dead or alive) is made in the transcendent domain of the collective subconscious, a subconscious that is shared by Wigner, his friend, and all the rest of us.

There are linguistic clues. We might talk about MY body, MY brain, MY mind, thus indicating that we do not identify the essential 'I' with any of these. One does not say, 'my' consciousness, however. It just doesn't feel right. The Sanskrit language of ancient India has many different words for 'consciousness'. The common everyday word is 'samjñā'. The prefix 'sam' means 'together'; 'jñā' is the root for 'to know'. Thus samjñā has the connotation of 'knowing together'.

In his Pulitzer Prize-winning book, *Gödel, Escher, Bach*, Douglas Hofstadter discusses what he calls *tangled hierarchies* or strange loops. As a mathematical example of a strange loop, consider the deceptively simple looking equation

$$x = -1/x$$

Nowhere in the world of real numbers with which we are familiar is there a solution to this equation. If you plug +1 into the left-hand side and calculate, you will get -1 on the right. Similarly, -1 on the left will produce +1 on the right. The equation constitutes a complete mathematical system, but it displays an inconsistency. In order to resolve the dilemma, we have to step altogether out of the system of real numbers and define the imaginary number $i = \sqrt{-1}$. Then $i^2 = -1$ and, dividing both sides by i, we get $i = -1/i$, which satisfies our equation. We had to step out of the immanent level and onto an inviolate level in order to do it.

A verbal example of a strange loop is the so-called liar's paradox, which can be stated as follows:

This man is a fisherman who says, "All fishermen are liars."

Is he telling the truth, or is he lying? If he is telling the truth, then all fishermen are liars, and so he is lying. If he is lying, then not all fishermen are liars, and so he might be telling the truth. This paradox results in an infinite mad oscillation of 'liar - not liar' that goes on forever. The only way out is to refuse to play the game by stepping out of the loop altogether. Imagine what a computer would do if you fed it something like this. The poor thing would probably go into meltdown.

An example that displays clearly the tangled hierarchies might be the image of a movie projector on one wall projecting the image of a movie projector onto another wall, which image is projecting the first image. This is unmistakably a complete system, albeit an inconsistent one. Real movie projectors are capable of projecting, whereas images of projectors are not. Real projectors are on a higher rung of the hierarchical ladder, hence the paradox is due to an entangling of hierarchies, as the name suggests. In this case, the inviolate level is the existence of a third movie projector projecting the two images that constitute the tangled hierarchy. All tangled hierarchies, of necessity, must have an inviolate level on which the paradox is resolved.

The Dutch artist M. C. Escher was well known for his drawings depicting strange loops and tangled hierarchies. One of his works entitled *drawing hands* shows a left hand drawing a right hand, which is drawing the left hand. There the inviolate level is M. C. Escher himself, who drew them both.

In 1931, the mathematician Kurt Gödel proved that any complete mathematical (or logical) system is necessarily inconsistent and conversely, any consistent system is necessarily incomplete. This theorem has had a profound effect on mathematics, constituting as it does a sort of uncertainty principle for mathematical and logical systems. Gödel carried out his proof using the flawed logic of tangled hierarchies.

Well (I hear you saying), what's all this in aid of? What do tangled hierarchies and their underlying inviolate levels have to do with consciousness? If you think about the implications of quantum mechanics and the seemingly inescapable neurotic nature of our lives, it seems likely that the areas of our mental life of which we are commonly aware, including our sense consciousness of the (apparently) objective world and our subjective consciousness of our thoughts and feelings, constitutes a tangled hierarchy, complete in itself, but flawed with inconsistency. Our erroneous concept of ourselves as isolated entities is threatened by what we perceive as a fundamentally external, and hence potentially hostile world. In this way, we are driven to defend this false sense of self and from this defense, all of our neuroses and unhappiness arises.

The inviolate level, on which the inconsistencies are resolved, is the collective transcendent subconscious, a level of which we are only dimly aware, the level where not only magic, but also true sanity resides. The subconscious level also has a lower individual aspect, which is the storehouse of buried memories. It is also the level on which individual psychic powers are activated. The quantum psychokinesis that we discussed in Chapter Twelve is an example of such a power that is available to all of us. Another example that the parapsychologists seem to ignore and one that is particularly easy to learn is *dowsing*. A little old lady from Vermont taught me to dowse in an afternoon. The power does not, of course, reside in the forked willow branch or bent coat hangers at all. They simply provide the psychological lever, an excuse so that we don't have to face up to the implications suggesting that all of us humans have such powers available to us at any time.

It is on the collective level of the transcendent subconscious that the world of appearances is actualized and the myriad wave function collapses take place.

Morality resides on the higher collective level of the subconscious. Someone who has successfully penetrated only the individual subconscious level and has tapped into the psychic power available there will probably use this power for ego aggrandizement and may become, in effect, an evil sorcerer, because he still sees himself as isolated and hence threatened. On the other hand, someone who has penetrated the collective level of the subconscious has stepped into a resplendent realm and has gained the realization that all sentient beings are in this together and that our own happiness is inextricably bound up in the happiness of others. We all want to be happy; this is the drive that motivates the best and the worst of us. Our delusion tells us that happiness is a *zero-sum game*, however, and that if we are to become happy, it must necessarily be at someone else's expense.

Consciousness must be, in general terms, somewhat as I have sketched it. It is up to a cooperative effort between physics, psychology, philosophy, and religion to fill in the details.

CHAPTER SIXTEEN

A RAY OF SANITY

in which the scientific and spiritual disciplines enter into a mutually beneficial relationship.

Quite a few books have been written on the parallels and similarities between the teachings of Eastern religions and the nature of the world as revealed by quantum mechanics. None of these books reached any convincing conclusions as to why this should be so, however. I think that we may be able to fill this gap on the basis of the material that we have covered in the foregoing chapters. In essence, we have found scientific evidence that Man is much more than our current paradigm would allow or lead us to believe, and that we all participate in the creation of our universe on the collective transcendent level of the subconscious. Buddhist and Hindu yogis have reached an experiential realization of this level (and beyond) through meditation and other psychological techniques. Experience is far more useful than the conceptual map that we have been able to draw, and so I propose that we take full advantage of whatever the yogis are willing to tell us and try to learn from them. Their methods are as scientific as ours and they are fully deserving of our attention and consideration. In concentrating on Buddhist sources, I do not mean to imply a lack of respect for Western religions. On the contrary, I am convinced that the spectrum of our propensities and spiritual needs is broad and that a correspondingly broad range of religions is necessary in order to serve these needs. The chief distinction between Eastern religions such as Hinduism, Buddhism, and Taoism, and the Western religions (Christianity, Judaism, and Islam) is that the Eastern religions have a strongly esoteric bias, whereas the Western religions are almost exclusively exoteric in nature. For our purposes, more is to be learned in examining teachings based on a search for meaning within our selves than in a philosophical approach in which all truth is taken to reside in an omniscient Creator, forever separate from and superior to man, and whose Mind is essentially unknowable.

The goal of Buddhist practice is the realization of Śūnyatā or 'emptiness', as it is most often translated. Early Western scholars commonly mistook this as a search for nothingness, oblivion, and labeled Buddhism as nihilistic, a religion of death. They were wrong. The emptiness that Śūnyatā points to is the emptiness of own-being, that is, independent existence.

At the first stage, known as the Śrāvaka vehicle, the object is to realize the emptiness, or lack of self, in the person. This is essential, because it is clinging to the idea that one is a single, permanent, and independent self that is the cause of all one's suffering. Please understand that the personality or ego as such is not the problem. These are obviously ever-changing and composite and do not correspond to the problematic self, which is really only a negative emotional response, and nothing more.

We argued in the last chapter that in some sense our ongoing stream of subjective and objective consciousness feels like a self. We noted that we never refer to it as *my* consciousness. But clearly, consciousness is a chain of momentary phenomena and cannot qualify as the sort of self that we defend so fervently. The spiritual (as opposed to secular) Sanskrit word for consciousness is vijñāna. The prefix *vi* means partial or divided; *jñāna* translates as knowledge. Thus consciousness constitutes a *partial knowledge*, implying that there is something more. The transcendent subconsciousness is referred to as the ālayavijñāna, or storehouse consciousness. Sanskrit, as you can see, is very descriptive.

However we look for a single, permanent, independent self, we fail to find one. All that we find is a vague concept projected from time to time onto our field of experiences. Any self that might qualify as the deserving object of our intense emotional reactions is clearly non-existent in absolute terms. The establishment of this experience is the goal of the first turning of the Dharmacakra, the wheel of the dharma or teachings.

In the third century CE, there lived in India a man named Nāgārjuna. His teachings, built on the foundation of the śrāvaka vehicle, were known as the Mādhyamaka. They were intended to establish the emptiness of the self-nature of all phenomena. The mind manifests as a stream of subjective and objective consciousness, along with the objects of consciousness. This is acknowledged to be *relative truth*, relative because each element of the stream arises only in dependence on the other, with neither the two aspects of consciousness nor their objects having a self-nature of its own. The Mādhyamaka teaching is a tough teaching and works to frustrate the reasoning mind into giving up its preconceived ideas concerning the nature of the world. In this sense, it is rather like a Zen koan. Nāgārjuna's method, for the most part, was to refute all philosophical positions and to offer none of his own. These teachings constitute what is known as the second turning of the Dharmacakra.

The third and final turning of the Dharmacakra consists of the teachings of Asanga, who lived and taught in India from the second half of the fourth century into the first half of the fifth century CE. Asanga had a thorough understanding of Nāgārjuna's teachings, but was aware that many people mistook the *negate everything* approach for nihilism. In order to clarify the situation, Asanga founded two schools: the Cittamātra (mind-only) school, which dealt with relative truth, and the Yogācāra school, which dealt with ultimate truth. Before we go on to examine these two schools and their teachings, I should define relative and ultimate truth.

The relative truth describes how things appear to consciousness in the light of analysis. The ultimate truth is the absolute nature of existence as established not only by accurate and minute analysis, but by actual experience. The relative truth is made up of concepts, which may be well-informed and highly reasoned concepts, but still are concepts. The ultimate truth is achieved by closely examining the relative truth to the point of achieving an experiential breakthrough. Physics operates on the relative level, a highly refined relative level, to be sure. With quantum mechanics, we have made an asymptotic approach to the ultimate truth, but the final realization must be accomplished by meditation and yogic techniques and not by the methods of physics. I will elaborate on this point further in the final chapter.

The śravaka vehicle is part of what is called the Hīnayāna, of which the Theravada sect of Śri Lanka is the outstanding example. The goal of the Hīnayānist is to gain liberation from his sufferings and to pass beyond the cycle of birth, sickness, old age, death, and rebirth. The Mayāyāna schools, of which the Mādhyamaka, the Cittamātra, and the Yogācāra are a part, have as their goal the much more ambitious task of leading all sentient beings to full Enlightenment.

At the śravaka stage, it is taken for granted that there is a truly existent world out there beyond the senses (the Hīnayānaists believed in the materiality of atoms). The Cittamātrins questioned this, as we have much more recently, and came to the conclusion that the nature of existence is described by monistic idealism. Where we have been led to this understanding by scientific reasoning, the Cittamātrins have arrived at this point by direct experience. They saw that the division of each moment of awareness into an inner perceiving consciousness and an outer perceived object is a purely conceptual invention. In a dream, one experiences inner perceiving consciousness and outer perceived objects, however, when one wakes the entire drama is realized to be a mental event, which Asanga characterized as *illusory truth*. Thus the appearance of objects of perception is no proof that they exist independent of the mind. In this way, the duality of mind and matter is erased. This suggests quite strongly that there are no limits to the power of the mind, hence it is perfectly reasonable to expect that we can reach full Enlightenment as the Buddha did and then work effectively for the enlightenment of all beings. This is really the only worthwhile activity by whatever route you wish to proceed.

We have been led to much the same conceptual understanding by our inquiries in physics. The path taken by the physicist and the path taken by the Buddhist Cittamātrin are complementary and have both led to the same conclusion. I should point out, however, that with us, this amounts to knowledge.

The Cittāmatrin, however, has reached this conclusion as an experiential realization, which gives it the strength of wisdom.

Can our knowledge (jñāna) be converted to wisdom (prajñā)? Yes, it can. That will be the subject of the last chapter.

I mentioned dreams, and in the case of the Cittamātra, it is particularly useful to pursue this topic further. Let's start out by considering the question: how do you know that you are not dreaming right now? You might answer by saying that dreams are never as vivid as the experiences you are having now; the colors are never so bright and the forms are not so clear and precise. Someone who has experienced *lucid dreaming* (where one carries one's waking consciousness intact into the dream state) might argue this with you and tell you that in his experience, the dream state can be just as vivid as the so-called waking state. There are yogic practices for inducing such lucid dream states.

Many people, even in our culture, have experienced them spontaneously. Your most convincing argument to yourself might be that the waking state has continuity and a fairly predictable pattern of events, whereas dreams are unpredictable and can change in bizarre ways with no warning. But dreams can stabilize and it is not unusual to have serial dreams night after night where events pick up where they left off the night before and considerable continuity is established. I might make the observation that waking reality can also become pretty bizarre.

The bottom line is that there is no characteristic of waking experience that clearly distinguishes it from dreaming. All arguments to the contrary are only a question of degree and of one's emotional predisposition. We want to believe that we are awake, because we want to believe that the world is solid and predictable and that we are in control. This need to protect our fictitious singular, permanent, and independent self explains why we solidify the waking reality as we do and accord it absolute status.

The Cittamātrins explain the phenomena of dreaming as a dissolving of the immanent level of consciousness back into the subconsciousness (ālayavijñāna) like waves in an ocean. The immanent level then taps into material there and creates images that the mind takes to be real and experiences as waking reality. The Cittamātrins are not saying that there is no difference at all between waking and dreaming experiences. What they are saying is that the difference is not an essential one.

The other argument that we have to deal with is that in waking reality there seems to be a consensus as to what the world is like, whereas in dreams this consensus seems to be lacking. Consensus, however, is also a matter of degree. We have no reason to believe that any of us sees or experiences the world in exactly the same way as anyone else. In fact, there is considerable evidence to the contrary. We know that there is a certain commonality of view with members of our own social, racial, and economic group, but even there we can get an argument. A stone-age Indian in the outback of Brazil does not view the world in anything like the same way that a Northern European does. For us, air is something to breathe and water can drown us. With fish, it is just the other way around. So much for consensus.

Let's proceed now to a brief discussion of the Yogācāra school, which is concerned exclusively with ultimate truth. The Yogācārans take issue with the eternal negations of the Mādhyamaka. They argue that if the mere absence of conceptual contrivance constituted ultimate truth, then one would have nothingness, not even empty space. How could such a total void possibly account for the manifestations of samsāra and nirvāna, which appear as vivid impure and pure manifestations respectively? The Yogācārans cannot hold that consciousness on any of its levels is responsible because they know that consciousness is ultimately without any self-nature. There must be something that is totally beyond the conceptualizing process. This they call Wisdom Mind (for want of a better name). It is the only entity that has absolute and ultimately true existence. It is from this viewpoint that all of the Buddhist Tantras proceed. Their goal is the realization of Wisdom Mind. What more can we say about Wisdom Mind? Not much.

At the beginning of this chapter I implied that physics had something to contribute to religion and it is this.

When the great Buddhist teachers were driven out of Tibet by the Chinese in the 1950s, many of them settled in the West and began to teach things that had been very closely held, even from Tibetan monks. The choice, as they realized, was either to propagate the Vajrayāna (or diamond vehicle) of tantric Buddhism widely and without reserve, or risk losing it altogether. The Tibetans are very traditional people, having lived in isolation on their plateau behind the wall of the Himālayas for thousands of years. Their teaching analogies often fall strangely on the Western ear. The Buddhist path is a two-pronged one of Wisdom and Compassion. The first step in gaining wisdom is to dispel primitive notions about the nature of existence. The view of physics and the view of the Cittamātra school concerning the nature of relative reality are in full and complete agreement. Thus physics is in the position to do religion a service by providing a thoroughly Western rationale for the Cittamātra teachings. Perhaps religion will return the favor by teaching physicists how to meditate.

CHAPTER SEVENTEEN

WORDS CORRUPT

You might well ask, as I have, the question: how did we get ourselves into such a delusional state and what is it that keeps us here. After all, physics and Buddhism both uncovered the error (2500 years ago in the case of Buddhism) and are able to make a convincing argument for the nature of existence being other than our paradigm would have us believe. Moreover, the materialistic paradigm is, to say the least, not in our best interests. It isolates us from meaningful relationships. It burdens us with cruel and painful neuroses. It reduces us to mere machines made of smelly meat, without any free will of our own. How did this hideous state of affairs that the Indian spiritual traditions call samsāra come about and why do we cling to it so? Is God punishing us for some original sin? Are we punishing ourselves for some imagined guilt? Is all of this just some cosmic bad joke?

First of all, the problem is not an exclusively human one. It is common to all sentient beings. If you examine the behavior of animals, you will see much the same pattern of neurosis everywhere. All sentient beings instinctively defend a non-existent permanent, independent, and singular self. The difference between animals and humans is that animals live in monistic ignorance, thoroughly embedded in nature, victims of stupidity and fear. They have no ego or intellect, which should tell you that ego is not the problem, as it is often accused of being.

Our human delusions, on the other hand, are dualistic in nature: me in here, other out there. We are aware of our individuality, and we know in our rational minds that the ego and the personality are dependent, impermanent, and that (in some sense) consciousness is shared, but we continue to go through these ugly knee-jerk reactions like puppets whose strings are being pulled. The Buddhists tell us that samsāra has no beginning. Clearly, though, the duality arose at some point in our development. Where did it come from? It came from language, the spoken word.

The subject-object split is inherent in language. There is always a subject and an object connected by a verb. Even in the most primitive languages where one might point at an object and name it, for example, 'drum', there is always the implied phrase, 'this is a drum', which excludes the possibility of it being 'not drum'. You simply cannot get away from the duality inherent in language.

In Chapter One, I expressed the opinion (by no means universally accepted) that language emerged to fulfill a protective role. If a threat could be named, then its power to harm would be defused to some extent. This might mean identifying what you thought was a snake as a vine or it might mean categorizing a serious threat and then using our intellect and experience to defeat it. Either way, language is the key. It was this need, and not the need to communicate, that caused language to appear. I am reasonably certain that our remote ancestors were communicating quite adequately by telepathic images long before language appeared in the world. The higher mammals probably communicate in this way as well. I have seen some evidence of telepathic ability in cats. You probably have too, if you stop and think about it.

The late prelingual phase of Man's development must have, in many ways, been like a bad acid trip as the pressure that was to bring forth language increased to a point of criticality. Language undoubtedly has the power to calm uncontrolled chaotic hallucinations, but there is a price. Language was the key that enabled us to formulate abstract concepts and theories, but in so doing, it killed the fluid element in the world. It assassinated the magic. Language is the concrete that we pour on our concept of the world. It solidifies reality.

A. J. Korzybski, the founder of the general semantics movement, understood the two-edged nature of language very well. He noted that the primary reason for the obvious lack of sanity in the world lay in the fact that people confuse their ideas about the world (as expressed in language) with the way the world really is. Korzybski claimed, and I think that he was right, as far as it goes, that most people confuse the word, which is only a symbol, with the thing that it represents. The thing itself is luminous, magical, and resplendent. The word is not. Think of the beauty of a sunset and how it makes you feel for the brief moment during which you can silence the chatterbox in your head. The word 'sunset' contains none of the reality of the genuine article. Poetry is man's attempt to describe something in words and, at the same time, to evoke as much of the magic and beauty as is humanly possible. In this sense, poetry is the highest literary form. The point is that whatever words you use, you must be aware that the words are merely abstractions and do not, in any way, reflect the structure of the thing itself. The event level is quite unspeakable. That's right, ultimate truth is totally beyond any verbal description. Once you say it is 'this', then you exclude the possibility that it is also 'that'. We have seen that quantum mechanics, which, in some ways is a bridge between the relative and ultimate levels, is nothing but probabilities and possibilities. The total openness that is ultimate truth refuses to be pinned down or confined in any way.

There is a nice little story in the Taoist tradition that makes the point: If one man asks about the Tao and another man tells him, then neither of them knows it.

Let's talk a bit about mathematics now. This too is a language, when you stop and think about it, formal and unambiguous, but still a language, conceived in the mind of man. Being a language, it has the same limitation that afflicts all other languages. It is inherently and inescapably dualistic. The very unambiguous nature of mathematics guarantees its duality. You simply can't get away from it. The fact that Schrödinger was able to write an equation to describe quantum mechanical wave functions tells you that these wave functions, expressed in mathematical terms, could not possibly correspond to the ultimate nature of existence. In essence, quantum mechanics describes the interaction between the collective and individual levels of the subconscious and the world of appearances. This is, perhaps (and only perhaps), the most subtle level of relative reality, but it is still relative reality. The ultimate reality, Wisdom Mind, or whatever other inadequate label you might chose, is totally beyond the descriptive powers of language, logic, or mathematics.

Eugene Wigner, the Hungarian physicist-philosopher, wrote an interesting essay in which he pointed out the unreasonable effectiveness of mathematics in the natural sciences. Now that we are aware that all levels of physical manifestation are our own creation (I have gotten that across, haven't I?), this should not be too surprising. It is just one more in the long list of clues that we have examined, all of which point to the monistic ideal nature of (relative) existence. This particular clue has been staring at us for a long time. I can remember wondering briefly about it in my undergraduate student days.

In elementary arithmetic, clearly direct observation led to the formulation of the arithmetic conclusion. If we have one apple on the table and then add another one, now there are two. Thus one plus one equals two. First the fact, and then the mathematical expression of it. There is nothing so profound about that. In higher fields of mathematics, which have been found to be useful in formulating physical theories, the observation did not lead to the mathematical conclusion. The mathematics had already been conceived as a playground for mathematicians, an area in which to demonstrate their ingenuity, and for no other purpose. Believe me, the mathematicians who invented complex numbers, analytical functions, tensors, matrices, and other such arcane mathematical objects, did so for their own amusement, not with any thought in mind that nature would actually be found to manifest in ways that would be describable in such terms.

Another startling example is group theory, as abstract a field of mathematics as you could possibly imagine. In the 1950s, Wigner pointed out that group theory was useful in describing the symmetries embodied in quantum mechanics. When Murray Gell-Mann and Yuval Ne'eman independently came up with the quark hypothesis, it was found to be an example of the special unitary group known as SU_3. It definitely appears, as Paul Dirac once suggested, that God is a mathematician of a very high order.

It would be best if we could all become completely enlightened and realize the Ultimate Truth. I would settle, in the short term, however, for trading in the miserable paradigm that we presently live with for a paradigm based on the knowledge that we have discussed in this book. The question is, how to go about doing it.

CHAPTER EIGHTEEN

YOU CAN LEAD A HORSE TO WATER, BUT

In which we contemplate some radical
brain surgery.

In one of his poems written when the British Empire was at the height of its power, Rudyard Kipling told us that: "East is East and West is West, and never the twain shall meet". Well, Kipling is a long way from being politically correct these days, so perhaps I can presume to contradict him.

I have always been a firm believer in taking the best that you find and leaving the rest. Package deals should, in general, be avoided. They usually contain undesirable elements, otherwise the offering would not have been bundled in a package in the first place.

I have already expressed my opinion as to the relative advantages of East and West. The West have been the unquestioned leaders in science and technology. Western art and music have also been very inspiring, less so in the present than in the past to my way of thinking. The East has focused on the science of the mind in its philosophical and psychological aspects and has made much more progress in the transpersonal area than we have in the West, although I would be the first to admit that this has not produced a utopian society for them anymore than the Western approach has in the West.

I realize that not everyone will agree with these opinions, but I have been led to them by experience and observation and I stand by them. I once worked with a Chinese scientist who had defected from Communist China and had come to America. I remember very clearly an observation he made after being in America for a while. He said, "When you are young, America is Heaven and China is Hell, but when you are old, China is Heaven and America is Hell."

This is so because, despite the materialist rhetoric of the Chinese Communist government, the essential sanity of Chinese civilization has valued the acquisition of wisdom. They have seen this wisdom concentrated in their older citizens and have respected and revered them for it.

In the West, value is put upon earning ability, not wisdom. Older citizens beyond their (materially) productive years are a burden to be disrespected and warehoused in nursing homes until they die.

My Chinese friend's observation was quite accurate. The reason for this state of affairs is largely due to the overwhelming acceptance of the materialistic paradigm in the West and the stubborn survival of the traditional Buddhist and Taoist metaphysics of idealism in the East. An elderly member of a family with traditional values in China or India knows that his mind in the form of the ālayavijñāna is going to survive the death experience. He believes in *karma*, the law of psychological cause and effect that was once a part of Christianity until it was swept under the rug along with the Gnostics. He knows that the merit of his virtues will insure a fortunate rebirth in which he will be able to make progress toward a state of realization that will enable him to benefit all beings. This approach to life gives the elderly Easterner a serenity and air of sanity that others want to be around, and so his needs are cared for and he is lovingly regarded as a treasure to be cherished.

In the West, the game is very different. Elderly people are not well regarded and they often manage to live up to their reputations. Many, if not most elderly people in the West are concerned with what Ken Wilber has so accurately identified as the *Ātman Project*. In a nutshell, this is the death avoidance game and it takes truly bizarre forms. Many, if not most people in our society do not believe in their heart of hearts that they are going to survive death in any way whatsoever. As they grow older, they begin to make more and more unreasonable demands. Their instinct tells them that if they can exercise power, any power, even the power to annoy, this will somehow hold off death and oblivion, of which they are afraid. So old people get cranky and troublesome and their children shuffle them out of sight into a nursing home and then feel guilty for having done so. It is no-win all the way around.

With instant world-wide communication, a lot of Western materialism has spread East and taken root. On the other hand, some positive Eastern values have taken a tentative hold in the West. Our own Western science has now confirmed that the materialistic paradigm to whose music we have all danced, the paradigm that has had such a corrosive effect on our spiritual selves, is dead. Asanga and Schrödinger have joined hands in the vision that Sir Arthur Eddington realized in 1927. Total enlightenment is, admittedly, a lofty and daunting goal, but the acceptance of the metaphysics of monistic idealism, no longer a philosophical speculation, but an established scientific fact, is a reasonable and sane thing to do, and is a necessary first step on the path. So how do we go about it?

The Tibetan Buddhist high lamas, the Rinpoches (precious ones) make the thoroughly practical suggestion that one should first study the argument (as made in this book or any other that you might choose), then think about it until you have all of the questions settled in your mind. Finally, they say, meditate on it. By this, they do not mean that we should sit on the floor in the lotus position and start chanting mantras (although that is not a bad idea, either). What they mean is, put it into practice. Make the teaching a part of your life. Strive to see that all sentient beings are partners in this enterprise and that it behooves us to support each other. We all just want to be happy and the worst of us are only ignorant, nothing more. To think that another person is evil, is not a useful concept.

Everyone really is doing the best that they can and all of them deserve our compassion.

The Beatles said that; "All you need is love". What you really need is wisdom and compassion. I hope that this book has made some small contribution to you in the wisdom area.

Go for it! You have nothing to lose and an awful lot to gain.

Joseph Norwood

INDEX

A

Absolute space, 35
Space, 35
Acceleration, 14
Action at a distance, 14, 42
Adam and Eve, 5
Additive color mixing, 97
Aesthetic principle, 7, 118
Aether, 23, 35
Agriculture, 5
Alchemy, 13, 15
Alhazen, 8
Ampere, A.M., 18
Anthropic principle,
　final, 117
　participatory, 117
　strong, 116
　weak, 116
Arabic science, 8
Arago D.F.J., 18
Aristarchus, 8
Aristotle, 7, 8, 13
Asanga, 130-131
Aspect experiment, 107-108
Aspect, A., 107
Ātman, 3
Atom:
　Bohr model, 63
　plum pudding model, 52
　Rutherford model, 62

B

Barberini, Cardinal, 9
Barrow, I., 12
Beatles, 139
Becquerel, H., 52

Bell, J. S., 92, 103, 108-109
Bell's inequality, 104-106
Berkeley, G., 14, 36, 122
Bernoulli G., 15
Birkhoff, G., 96
Black holes,
　Kerr, 46
　Reissner-Nordstrøm, 46
　Schwarzschild, 46
Black plague, 6
Black, J., 15
Blackboy radiation, 52-54
Bohm, D., 92, 102-103
Bohr model of hydrogen atom,
63
Bohr, N., 62-63, 87
Bohr quantization relation, 63
Bondi, H., 48, 115
Boole, G., 94
Boolean logic, 94
Bootstrap, 120
Born, M., 68-69, 70, 72, 76
Bradley J., 17
Brahe, T., 11-12
Brownian motion, 54
Buddhism:
　Tibetan Vajrayāna, 3, 133
　Zen, 3

C

Calcite, 104
Calendar, 4
Caloric theory, 15-16
Calorimetry, 15
Cambridge University, 12
Capra, F., 120
Carnot, S., 16
Carter, B., 116

Cartesian dualism, 121
Cavendish, H., 17
Cavendish Laboratory, 52
Celsius, A., 15
Centrifugal force, 14, 34-36
Chardin, P. T. de, 117
Chew, G., 120
Cittamātra, 130
Clauser, J., 106
Clausius, R. J. E., 16
Clock, 26
Collective unconscious, 118
Color mixing:
    additive, 97-98
    subtractive, 97-98
Compassion, 139
Complementarity, principle, 88-89
Compton, A. H., 55, 59-60
Consciousness, 6, 78, 125-128, 130, 132
Conservation:
    of energy, 15-16
    of momentum, 11, 13
Contrafactual definiteness, 120
Copenhagen interpretation, 87-93
Copernicus, N., 8
Coriolis force, 14, 34
Correspondence principle, 89-90, 101
Cosmic background radiation, 48-49
Cosmological constant, 47-49
Cosmological principle, 48, 115
Cosmology,
    big bang, 48-49
    geocentric, 7, 12
    heliocentric, 8
    steady state, 48
Coulomb, C., 17
Creationist, 5

Curie, M., 52
Curie, P., 52
Curvature of space, 40-41

D

Dark matter, 49
Darwinian evolution, 5
Davisson, C. J., 64-66
Davy, H., 15
De Broglie, L. V., 60-61, 63-64, 92
Degenracy pressure, 76
Delayed choice experiment, 83, 112
Descartes, R., 11, 13, 124
Determinism, vi-vii, 72, 76, 81
Dicke, R., 115
Dirac, P. A. M., 70-71
Dispersion, optical, 17
Distributive law of logic, 94
Dogma, 7
Doppler effect, 39, 47
Double-slit experiment, 81-86, 88
Dreams, 132
Du Fay, 17

E

Eddington, A. S., 44, 47, 123
Ego, 2, 4
Einstein, A., 25, 54, 58, 87, 102
Electromagnetic waves, 21-22
Electromagnetism, 18
Electron, 52, 55
Energy:
    conservation of, 15-16
    in relativistic, 30-31
Eötvös experiment, 36, 38

Eötvös, R. v., 36
EPR, 102
Equivalence principle, 37-41
Escher, M. C., 127
Eternalism, 4
Euler, L., 15
Everett, H., 79
Evolution, Darwinism, 5
Experimenter effect, 110

F

Fahrenheit, G. D., 15
Faraday, M., 18-20
Fermat's principle, 42
Fitzgerald contraction, 27
Fitzgerald, G. F., 25-26
Fixed stars, 35
Fluxions, 13
Force, 13
Foucault, J. B. L., 18
Fourier analysis, 75
Frank, P., 123
Franklin, B., 17
Freedom, S., 106
Fresnel, A. J., 18

G

Galilean telescope, 9
Galilei, G., 8-11, 13, 124
Galileo, falling bodies, 10
Galvanic, L., 18
Gamow, G., 49
Gell-Mann, 32
Geocentric cosmology, 7, 12
Geodesics, 42
Gerlach, W., 95
Germer, L. H., 64-66
Gnostic Christians, 3

God, 3
Gödel, K., 127
Gold, T., 48, 115
Gravitational lens, 44, 85
Gravitational spectral shift, 38-41
Gravitational waves, 45
Gravity:
    Einsteinian, 34-50
    Galilean, 10, 13
    Newtonian, 13

H

Halley, E., 13-14
Hamilton, W. R., 70
Harrison, E. R., 47
Hawking, S., 13
Heisenberg microscope, 73
Heisenberg, W., 68-70, 72
Heliocentric cosmology, 8
Helmholtz, L., 16
Henry, J., 19
Hertz, H., 21, 55
Hidden variables, 92, 103
Hīnayāna, 131
Hofstadter, D., 126
Holmes, S., 11, 78
Hooke, R., 13
Hoyle, F., 48, 115-117
Hubble, E., 47
Huygens, C., 11, 13

I

Idealism, 121-123
Illusion, vi
Illusory truth, 131
Inertia, 34-36
Inertia, principle of, 10, 13

Inertial frame, 14, 22, 34
Interference, wave, 23
Islam, 3

J

Jeans, J., 53
Johnson, S., 122
Jordan, P., 69-70
Joule, J. P., 16
Judaism, 3

K

Karma, 6
Kelvin, W. T., 16
Kepler, J., 11-12
Kepler, laws of motion, 12, 14
Kipling, R., 137
Klein-Gordon equation, 71
Knowledge, v
Korzybski, A. J., 135

L

Lagrange, J. L., 15
Language, 1-2, 5, 134-136
Lattice, logic, 97-100
Lemaitre, G., 47
Lenard, P., 55
Light, 18
Light, speed of, 22
Lippershey, 9
Locality, 32, 93
Locke, J., 13, 124
Logic:
    Boolean, 94
    quantum, 94-101
Logical positivism, 68
Lorentz, H. A., 25-26

M

Mach, E., 14, 36, 52, 69-70
Mach's principle, 36, 42
Mādyamaka, 130
Magnetism, 18
Many-worlds, 79, 121
Marconi, G., 21
Margenau, H., 93
Mathematics, 135-136
Matrix mechanics, 71
Maxwell, J. C., 17, 20-21
Mayāyāna, 131
Mechanics, vi
Medici, Cosimo di, 9
Medieval period, 6
Michelson, A. A., 23
Michelson-Morley experiment, 23-24
Millikan, R. A., 56, 58
Mind-stuff, 123
Minkowski diagram, 29-30, 32-33
Minkowski, H., 29
Minkowski space, 29-31, 41
Momentum:
    conservation of, 11, 13
    relativistic, 30-31
Morley, E. W., 24
Morse, S. F. B., 20
Mössbauer effect, 40

N

Nāgārjuna, 130
Neumann, J. v., 92-93, 96
Newton, I., 11-15, 17, 21
Nihilism, 4
Nirvāna, 133

Non-locality, 86

O

Objectivity, vi, 81
Ockham's razor, 125
Oersted, H. C., 18
Ohm, G. S., 18
Olbers' paradox, 46-47
Omega point, 117
Optics, 17

P

Padua, 8
Pagels, H., 118
Paradigm shift, 1
Paranoia, 4
Parapsychology, 110
Pauli, W., 70
Penzias, A., 49
Perfect cosmological principle, 48, 115
Perrin, J., 54
Philosophy, vi, 4, 78
Photoelectric effect, 54-59
Photon, 58-59
Pisa, 8
Planck, M., 53-55
Planck's constant, 54
Poincaré, H., 25-26
Polarization, 99-101
Pope Urban VIII, 9-10
Pound, R. V., 40
Priest, 3, 5
Principia, 13
Protagoras, 119
Psychic forces, 1, 3
Psychological lever, 111
Psychology, vii, 124-128

Psychotropic drugs, 2
Ptolemy, 8

Q

Qabbalism, 3
Quantization relation, Bohr, 63
Quantum logic, 94-101
Quantum measurement, 77-81
Quantum numbers, 89
Quantum psychokinesis, 110-113

R

Rayleigh-Jeans law, 53
Rayleigh, J. W. S., 53
Rebka, G. A., 40
Reincarnation, 9
Relative truth, 130
Religion:
    esoteric, vii, 3, 6, 129-133
    exoteric, vii, 3, 5, 129
    primitive, 1
Renaissance, 6
Rgveda, 3
Röntgen, W. K., 51
Royal Society, 13
Rumford, B. T., 15
Rutherford, E., 61

S

Sagredo, 9
Saliviati, 9
Samsāra, 133-134
Sanskrit, 126, 130
Scalar, 43
Schmidt, H., 110, 112

Schrödinger equation, 71-72
Schrödinger, E., 70-72
Schrödinger's cat, 80-81,
   125-126
Schwarzschild, K., 43
Scientific method, vii, 7, 24
Selleri, F., 109
Shaman, 3, 5
Simplicio, 9
Simultaneous events, 29-30
Solipsism, 79-80
Sommerfeld, A., 68
Space-time, 29
Spin, 94-96
Śrāvaka, 130
Statistical   ensamble
   interpretation, 81
Stern, O., 95
Strange loops, 126-127
Strings, 121
Subconsciousness, 125
Subtractive color mixing, 97
Sufism, 3
Śūnyata, 129
Superluminal signals, 32-33
Superposition, principle of, 10,
   11

T

Tachyons, 31-32
Tangled hierarchies, 126-127
Tarozzi, G., 109
Teleology, 116
Telescope:
   Galilean, 9
   Newtonian, 17
Tensor, 43
Thomson, G. P., 66
Thomson, J. J., 52, 55

Three-polarizer paradox, 100-101
Time dilation:
   special relativistic, 26-27
   gravitational, 38-41
Tribal identity, 2

U

Ultimate truth, 131
Ultraviolet catastrophe, 53
Uncertainty principle, 74-76
Urban VIII, Pope, 9-10

V

Vajrayāna Buddhism, 3
Vector, 43
Vedic tradition, 4
Venice, 9
Volta, A., 18

W

Wave interference, 23
Wave mechanics, 71
Waves:
   electromagnetic, 21
   gravitational, 45
   light, 18
   sound, 22
Weyl, H., 71
Wheeler, J. A., 83, 85, 117
Wien, M., 53
Wigner, E. P., 78, 125, 136
Wilson, R., 49
Wisdom, v, 139
Wittgenstein, L., 68
Wren, C., 13-14

X

X-rays, 51

Y

Yogācāra, 131
Young, T., 18, 81

Z

Zen Buddhism, 3

ABOUT THE AUTHOR

My qualifications for writing a book that spans physics, philosophy, psychology, and Eastern religions include a Ph.D. in Physics, received in 1967. I worked with NASA from 1957 to 1967 and held an Assistant Professorship at the University of Miami from 1967 to 1972, and an Associate Professorship at East Carolina University from 1977 to 1983. I was active in Star Wars research from 1984 to 1986 and carried out studies for the design of an entirely benign nuclear reactor from 1986 to 1989. The latter experience provided the inspiration for my novel Uttermost South. In all those years, my thoughts have as often been of "why" as well as the more traditional "how". I thursted for nothing less than knowledge of the nature of existence.

I take a genuine joy in intellectual inquiry in general and in physics in particular. It is not in my nature to let fashions or politics place limits on what interests me or in what I choose to investigate. Consequently, my career has not exibited a smooth upward progression, however I have pursued the goal of educating myself relentlessly and I have no regrets.

On the spiritual side, I have been a student and a practitioner of Tibetan Vajrayana Buddhism since 1980. 1 would not ordinarily mention such a private matter in an author biography, but in this case it has provided the key that has enabled me to have confidence in the material presented in Physics, Consciousness, and the Nature of Existence. I am the author of seven books and about 50 journal articles, including two novels, Flight of the Falcon and Uttermost South, and two books on yacht design.

www.ingramcontent.com/pod-product-compliance
Lightning Source LLC
Chambersburg PA
CBHW031121180526
45160CB00005B/44/J